DIGITAL RADIOGRAPHY IN PRACTICE

DIGITAL RADIOGRAPHY IN PRACTICE

By

QUINN B. CARROLL, M.Ed., R.T. (R)

CHARLES C THOMAS • PUBLISHER, LTD.
Springfield • Illinois • U.S.A.

Published and Distributed Throughout the World by

CHARLES C THOMAS • PUBLISHER, LTD.
2600 South First Street
Springfield, Illinois 62704

© 2019 by CHARLES C THOMAS • PUBLISHER, LTD.

ISBN 978-0-398-09271-9 (paper)
ISBN 978-0-398-09272-6 (ebook)

Library of Congress Catalog Card Number: 2019002485 (paper)
2019003074 (ebook)

With THOMAS BOOKS *careful attention is given to all details of manufacturing
and design. It is the Publisher's desire to present books that are satisfactory as to their
physical qualities and artistic possibilities and appropriate for their particular use.*
THOMAS BOOKS *will be true to those laws of quality that assure a good name
and good will.*

Printed in the United States of America
MM-S-2

Library of Congress Cataloging-in-Publication Data

Names: Carroll, Quinn B., author.
Title: Digital radiography in practice | by Quinn B. Carroll.
Description: Springfield, Illinois, U.S.A. : Charles C Thomas,
    Publisher, LTD., [2019] | Includes bibliographical references and
    index.
Identifiers: LCCN 2019002485 (print) | LCCN 2019003074 (ebook) |
    ISBN 9780398092726 (ebook) | ISBN 9780398092719 (paper)
Subjects: | MESH: Radiographic Image Enhancement | Image
    Processing, Computer-Assisted—methods | Radiographic Image
    Interpretation, Computer-Assisted | Technology, Radiologic
Classification: LCC RC78.4 (ebook) | LCC RC78.4 (print) | NLM WN
    26.5 | DDC 616.07/572—dc23
LC record available at https://lccn.loc.gov/2019002485

REVIEWERS

Wayne McKenna, RTR, ACR, CAE

Program Coordinator, School of Radiography
University of Prince Edward Island
Charlottetown, PEI, Canada

Patrick Patterson, MS, RT (R) (N), CNMT

Director, Radiography Program
State College of Florida, Manatee-Sarasota
Bradenton, Florida

Robert L. Grossman, MS, RT (R) (CT)

Instructor, Radiography
Middlesex Community College
Middletown, Connecticut

Donna Endicott, MEd, RT (R)

Director, Radiologic Technology
Xavier University
Cincinnati, Ohio

PREFACE

This work is intended to provide medical radiography programs with an economical textbook that focuses on the practical aspects of digital radiography, limited in scope to information that will be pertinent to each graduating student as he or she enters into clinical practice. Nearly all textbooks to date claiming the title "digital radiography" have dealt primarily with the managerial aspects of the topic at the expense of any practical information on how digital imaging actually works and its clinical implications for the daily practice of radiography.

Since no other books have yet filled this need, much of this information has originated primarily from direct contact by the author with scientists at Philips Healthcare, FujiMed and CareStream Health (previously Kodak), who were directly involved in actually developing these technologies, in addition to numerous "white papers" published by companies that produce digital radiography equipment. These sources are all listed at the end of the book in the References.

A much more extensive treatment of the subject is found in *Radiography in the Digital Age* (Charles C Thomas, Publisher, Ltd., 2018), a work by this author of over 900 pages that provides experimental evidence and in-depth explanations for all the various aspects of digital technology of interest to the radiographer, in addition to the underlying physics of radiography, principles of exposure and technique, and a thorough coverage of radiation biology and protection. Many of the lucid illustrations in this textbook are borrowed here to make digital radiography comprehensible to the student, but in this textbook we focus only on digital topics and state the facts with such brief explanatory material as each topic will allow.

Use of the *glossary* is highly recommended whenever a concise definition is needed for a particular term.

The goal of the author is to provide an accurate and adequate description of all of the aspects of digital images and digital equipment, and their implications for radiographic technique and clinical application, but to do so in the most student-friendly way possible by providing crisp, clear illustrations along with readable text. Many digital topics are intimidating, and every attempt is made to reduce these topics to a descriptive, non-mathematical level that can be intuitively understood by the average student. Feedback from educators and students is welcome.

Ancillary Resources

Instructor Resources CD for Digital Radiography in Practice: This disc includes hundreds of multiple-choice questions *with permission* for instructors' use. **Answer keys** for all chapter-end questions in the textbook are included, along with keys to the multiple-choice question banks. (Instructors desiring laboratory exercises and more extensive question banks are encouraged to purchase the *Instructor Resources CD for Radiography in the Digital Age,* also available from Charles C Thomas, Publisher, Ltd.) The website is ccthomas.com.

PowerPoint™ Slides for Digital Radiography in Practice: PowerPoint™ slides are available on DVD for classroom use. These are high-quality slides with large text, covering every chapter of the textbook. (Instructors desiring more extensive slides are encouraged to purchase the *PowerPoint™ Slides for Radiography in the Digital Age,* also available from Charles C Thomas, Publisher, Ltd.) The website is ccthomas.com.

Student Workbook for Digital Radiography in Practice: This classroom supplement is correlated with the PowerPoint™ slide series for in-classroom use. Although it can be used for homework assignments, it is designed to deliberately provoke student participation in classroom instruction while avoiding excessive note-taking. All questions are in "fill-in-the-blank" format, focusing on key words that correlate perfectly with the slide series. Available from Charles C Thomas, Publisher, Ltd. The website is ccthomas.com.

ACKNOWLEDGMENTS

Many thanks to Wayne McKenna, Donna Endicott, Patrick Patterson, and Bob Grossman for reviewing this textbook. The assistance of Dr. Ralph Koenker at Philips Healthcare, Gregg Cretella at FujiMed, and Dr. Daniel Sandoval at the University of New Mexico Health Sciences Center is also greatly appreciated.

I am grateful for the perpetual support of family, friends, and prior students who make all the effort meaningful. I dedicate this work to practicing radiographers everywhere, on the front-lines of patient care.

CONTENTS

DIGITAL RADIOGRAPHY IN PRACTICE

- analog is a film

Chapter 1

NATURE OF THE DIGITAL RADIOGRAPH

■ ■

Objectives

Upon completion of this chapter, you should be able to:

1. Analyze the differences between analog and digital data and how they relate to radiographic images.
2. Define the three steps in digitizing any analog image.
3. Explain the relationships between bit depth, dynamic range and image gray scale.
4. Describe the aspects of a digital image matrix and how it impacts image sharpness.
5. Define *voxels, dexels (dels),* and *pixels* and distinguish between them.
6. Describe the nature of *voxels* and how the x-ray attenuation coefficient for each is translated into the gray levels of pixels.

Development of Digital Radiography

The first application of digital technology to radiographic imaging occurred in 1979 when an analog-to-digital converter was attached to the TV camera tube of a fluoroscopy unit. It makes sense that digital conversion would first occur with fluoroscopy rather than "still" radiography, because the signal coming from a TV camera tube was in the form of electrical current rather than a chemical photo-

graphic image, and computers are based on electricity.

Three years later, in 1982, the introduction of digital picture archiving and communication systems (PACS) revolutionized the access, storage and management of radiographic images. Coupled with *teleradiology,* the ability to send electronic images almost instantly anywhere in the world, the efficiency of medical imaging departments in providing patient care and diagnoses was also revolutionized.

Computed radiography or "CR" became commercially available in the early 1980s, but was at first fraught with technical problems. It was found that "screens" coated with certain fluorescent materials, which had been used to convert x-ray energy into light that exposed films, could be made to glow a *second time* afterward when stimulated by laser beams. This stimulated light emission, using only the *residual energy* remaining in the screen after x-ray exposure, was very dim indeed. But, after being captured by light-sensitive diodes and converted into a meager electrical current, it could be electronically and digitally amplified before being processed by a computer to produce a bright radiographic image on a display monitor.

Early CR systems required an approximate doubling of x-ray technique, resulting in a doubling of patient exposure. But they were later refined and paved the way for the development of *digital radiography* or "DR," first demonstrated in 1996.

The advancing miniaturization of electronics finally led to x-ray detectors that are smaller than the human eye can detect at normal reading distance. By constructing image receptor "plates" with thousands of these small detectors laid out in an *active matrix array,* it was possible to convert the latent image carried by the remnant x-ray beam *directly* into electrical current, called *direct-conversion DR.* Indirect-conversion DR units use a phosphor plate to first convert the x-rays into light, then the active matrix array converts the light into electricity. Direct-conversion systems convert the x-ray energy directly into electricity without the intermediate step of converting x-rays into light. Indirect conversion units have the advantage of saving patient radiation dose, but direct-conversion units produce better resolution. Since these are both desirable outcomes, both types of systems continue in use.

All CR and DR imaging systems ultimately produce an *electronic signal* that represents the original image information. It is this electrical signal that is "fed" into a computer for digital processing and then finally displayed on an electronic display monitor. Although both CR and DR systems continue in use, after more than two decades of refinement DR has emerged as the state-of-the-art technology for medical radiography.

Nature of the Digital Image

DR, CR, DF (digital fluoroscopy), digital photography, and all other methods of acquiring a digital image result in the creation of a *matrix* of numerical values that can be stored in computer memory. A matrix is a pattern of cells or locations laid out in rows and columns as shown in Figure 1.1. Each location or cell can be identified by its row and column designations, which the computer keeps track of throughout any processing operations. Each location or cell in the matrix is referred to as a *pixel,* a contraction of the term *picture element.* Each pixel in an image is assigned a single numerical value, the *pixel value.* For radiographs, the pixel value represents the brightness (or darkness) assigned to the pixel's location in the image. This brightness level is taken from a range of values stored in the computer that represent different shades

COLUMNS

	A	B	C	D	E	F	G
1	A1	B1	C1	D1	E1	F1	G1
2	A2	B2	C2	D2	E2	F2	G2
3	A3	B3	C3	D3	E3	F3	G3
4	A4	B4	C4	D4	E4	F4	G4
5	A5	B5	C5	D5	E5	F5	G5
6	A6	B6	C6	D6	E6	F6	G6
7	A7	B7	C7	D7	E7	F7	G7

ROWS

Figure 1-1. A digital image matrix with the location of each cell designated by column and row.

from "pitch black" all the way to "blank white," with hundreds of shades of gray in between.

Light images enter through the lens of a camera in *analog* form, that is, the various intensities of light can have any value. Likewise, x-rays from a radiographic projection enter the image receptor plate in analog form. During a medical sonogram procedure, sound waves enter the transducer in analog form, as do radio waves emanating from the patient during an MRI scan. All of these forms of input must be converted into *digital* form so that we can manipulate the resulting images as we wish to do.

To better distinguish between analog and digital data, imagine that you are standing on a railroad track (preferably with no trains coming) as shown in Figure 1-2. You can choose to walk along the metal rails, doing a balancing act. Or, you can choose to hop along the wooden cross-beams, stepping from tie to tie. The metal rails are *continuous*, consisting of smooth, unbroken lines. Your progress along the rails can be measured in *any fraction* of distance—meters, millimeters, microns—there is no limit to how many times you can divide these measurements into smaller and smaller units. An analog measurement can be as precise as we want, because, by using a continuous scale, it is *infinitely divisible.*

Now, suppose you choose to step along the cross-beams. The wooden ties are *discrete* or *separated* into distinct units. Your progress along them cannot be measured in fractions because of the spaces between them. You must count them in whole integers. Digitizing data *limits* the degree to which measurements can be subdivided. It also limits the *scale* from which measurements can be taken. For example, only so many railroad ties of a particular size can be laid between one point and another that is one kilometer away. In radiography, digitizing the pixel values limits the number of values that can exist between "pitch black" and "blank white." This makes them manageable, because there is not an *infinite number of values* to deal with.

For the purpose of building up an image and manipulating it, we need all pixel values to be *discrete*, that is, selected from a limited scale of pre-set values. If our scale is set from 0 (for blank white) to 4.0 (for pitch black), and we limit decimal places to the *thousandths*, then we will have 4000 values available to build up an image. This is more than enough to allow the image to be not only built up for initial display, but also to be "windowed" up and down, lighter

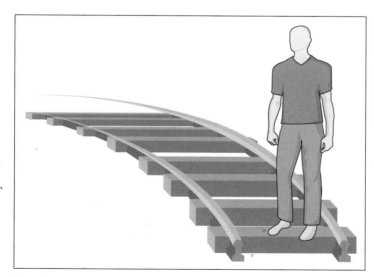

Figure 1-2. On a railroad track, the steel rails represent *analog* information—they are continuous and can be infinitely subdivided. On the other hand, the wooden ties represent *discrete* or *digital* information, since they cannot be subdivided into fractions as one steps from tie to tie. (From Q. B. Carroll, *Radiography in the Digital Age,* 3rd ed. Springfield, IL: Charles C Thomas, Publisher, Ltd., 2018. Reprinted by permission.)

or darker at will, across the entire range of human vision as needed. Yet, it is not an infinite range of values.

Mathematically, digitizing means *rounding* all measurements to the nearest available digital value. In the above example, an analog measurement of 1.0006 must be rounded up to the nearest thousandth or 1.001, a measurement of 1.00049 will be rounded down to 1.000. This rounding-out process may seem at first to be a disadvantage for digital computers. Strictly speaking, it is less accurate. However, when we take into consideration the limitations of the human eye and ear, we find that digitized information can actually be *more* accurate when *reading out* the measurement. This is why digital equipment is used to clock the winner of a race in the Olympics: you may not be able to *see* that the winning racer was just two-thousandths of a second ahead of the second-place racer, but a digital readout can make this distinction. *As long as the discrete units for a digital computer are smaller than a human can detect, digitizing the data improves readout accuracy.*

For digital photography and for digital radiography, if the units for pixel values are smaller than the human eye can detect, the resulting digital image will appear to have the same quality as an analog photograph or radiograph. Digitization of incoming analog data is the function of a device called the *analog-to-digital converter (ADC)*, which is used in all forms of medical digital imaging.

Digitizing the Analog Image

We can identify three basic steps to *digitizing* an image that apply to all forms of images. The **first step** is *scanning*, in which the field of the image is divided into a matrix of small cells. Each cell will become a *pixel* or picture element in the final image. In Figure 1-3*A*, the field is divided into 7 columns and 9 rows, resulting in a matrix size of 63 pixels. The photocopy scanner connected to your home computer can be heard making a pre-copying sweep before it makes the actual copy, performing this function of pixel allocation and matrix size determina-

tion. For computed radiography (CR), the processor or reader scans the exposed PSP plate in a predetermined number of lines (rows) and samplings (columns) that define the corresponding pixels.

For digital radiography (DR), the number of *available* pixels is determined at the detector plate by the number of hardware detector elements (*dexels*) physically built into the plate in rows and columns. In this case, *collimation* of the x-ray beam is analogous to the *scanning* function, because collimation effectively *selects* which of these detector elements will comprise the initial matrix of the latent image that will be fed into the computer for processing. A similar concept holds true for digital fluoroscopy (DF): The initial field of view (FOV) selected for a dynamic flat-panel system, or determined by the magnification mode of an image intensifier at the input phosphor, are analogous to collimation—they determine the matrix size, the pixel size, and the spatial resolution of the input image that will be processed by the computer.

All forms of digital imaging require the preliminary step of *formatting a matrix with a designated pixel size,* and whatever method is used, it would fall under the broad definition of *scanning.*

The **second step** in digitizing an image is *sampling,* defined as the detection and *measurement* of the intensity of signal coming into the system at each pixel location, Figure 1-3*B*. For standard photography, for CR, for indirect-conversion DR, and for CCD or CMOS cameras mounted atop a fluoroscopic image intensifier, this signal consists of the intensity of *light* striking each designated pixel area. For direct-conversion DR, the signal consists of the intensity of *x-rays* striking each pixel area. For an MRI machine, the signal consists of radio waves, and for sonography, sound waves. The type of imaging equipment being used determines the size and shape of the *aperture* or opening through which these signal measurements are taken. For example, the detector elements of a DR machine are essentially square, whereas the pixel aperture inside a CR reader is round, and the initial pixel samplings overlap each other, because a round laser beam is used to stimulate the PSP plate to

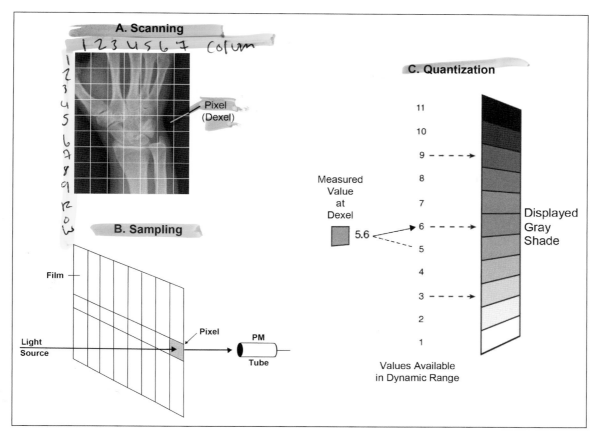

Figure 1-3. Three steps for digitizing an image: **A**) *Scanning* to format the image into a matrix of defined pixels (or dexels); **B**) *Sampling,* in which measurements are taken from each pixel or dexel; and **C**) *Quantizing,* in which each measurement is rounded to the nearest digital value available within the dynamic range.

glow, Figure 1-4. This overlapping effect must be "cropped" in order to form roughly square-shaped pixels for the final displayed image.

The third and final step in digitizing an image is *quantization.* In the previous section, we described how analog values must be effectively *rounded out* to form discrete values that the digital computer can recognize and manipulate. These values must be selected from a predetermined scale of numbers called the *dynamic range.* The dynamic range of any imaging system is the *range of pixel values, or shades of gray, made available by the combined hardware and software of the system to build up a final displayed image.* Actual values of the signal intensity measured, which will become the brightness level for every pixel, must each be rounded up or down by an *analog-to-digital converter (ADC)* to the nearest available gray

level in the preset dynamic range. In Figure 1-3**C**, there are only 11 such values available to choose from to build up this simplified image. This is the process of quantization or *quantizing the image.*

Bit Depth, Dynamic Range, and Gray Scale

The term *dynamic range* is frequently misapplied, even by physicists, and can be a source of confusion. For example, some have limited the term to describing the characteristics of a DR detector plate. But, with such a narrow definition, digital features such as *dynamic range compression* or *dynamic range control (DRC),* which alter the dynamic range during processing, would imply that we have effectively gone backward

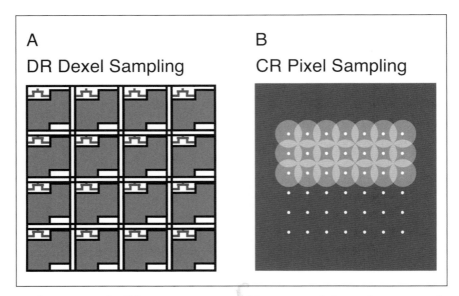

Figure 1-4. The *sampling aperture* for DR equipment is roughly square, ***A***. Since the aperture for CR is round, ***B*** (due to the round scanning laser beam in the CR reader), the original samplings must overlap each other in order to fill the *square* pixels of the final displayed image on the LCD. (From Q. B. Carroll, *Radiography in the Digital Age,* 3rd ed. Springfield, IL: Charles C Thomas, Publisher, Ltd., 2018. Reprinted by permission.)

in time to alter a characteristic of the image receptor plate during the initial exposure. The student must understand that although the terms *dynamic range* and *bit depth* are often used interchangeably in practice, we can find clarity by examining the *dominant* use of each term by experts.

The term *bit depth* is dominantly applied to *hardware devices such as computers, display monitors, and DR detector plates.* Bit depth is the *maximum range of pixel values the computer or other hardware device can store.* Bit depth is expressed as the exponent of the base 2 that yields the corresponding binary number; For example, if a typical DR detector has a bit depth of *10,* it can store 2^{10} = 1024 pixel values. A display monitor with a bit depth of *8* can store 2^8 = 256 pixel values (or shades of gray). We say that the monitor's pixels are "*8 bits deep.*" The bit depth of the human eye is only about *5,* or 2^5 = 32 different levels of brightness that it can discern. Since the bit depth of nearly all types of imaging equipment is well beyond the capacity of the human eye, the resulting images can be indistinguishable from analog images to us.

Therefore, the *full* bit depth of a hardware system need not be used in presenting images at the display screen, and doing so slows down computer processing time. Appropriate to the medical application of a particular imaging machine, the *system software* determines the range of brightness levels or pixel values that are made available for building up images at the display monitor. This is the *dynamic range—the range of pixel values, brightness, or gray levels made available by the combined hardware and software of an imaging system to build up a final image at the display monitor.*

The displayed brightness level at each pixel must be "selected" from this scale. This is the *quantizing* step in digitizing an image, Figure 1-3***C***. As with bit depth, the dynamic range of a system is usually expressed as a binary exponent, 2, 4, 8, 16, 32, 64, 128, 256, 528, 1024 or 2048. This range is available to *each* pixel in the image. Once the final image is built up and displayed, we define the *gray scale* of the image as the range of grays or brightness levels *actually displayed.*

To summarize, *bit depth* generally describes the capabilities of the hardware equipment being used to capture, process, and display an image. *Dynamic range* is the range of pixel values made available by the entire system (hardware and software) to build up a displayed image. And, *gray scale* is the range of grays actually apparent in the displayed image. Dynamic range is a *subset* of the bit depth of a system, and displayed gray scale is a *subset* of the dynamic range.

In medical imaging, there is an essential purpose for using a dynamic range that is much larger than the displayed gray scale. There must be "room" for image manipulation by the user, such as windowing the brightness or gray scale up and down. As shown in Figure 1-5, "levelling" or changing the window level of the image essentially slides the displayed gray scale up or down the available dynamic range. From an average brightness level, there must be room for several doublings or halvings of the displayed overall brightness. The same kind of flexibility must also be provided to allow for adjustments of the gray scale and contrast.

In addition to this, even more "room" must be allowed for special processing features such as image subtraction to be applied without "running out" of available pixel values, a phenomenon called *data clipping* (see Chapter 7). CT and MRI systems require a 12-bit dynamic range (4096 values) for their enhanced processing fea-

tures. Remember that the *main advantage* of digital imaging over conventional film imaging is its enhanced contrast resolution, which depends entirely on an extended dynamic range and the processing latitude it affords.

The "dynamic range" of the x-ray beam itself is said to be approximately 2^{10} or 1024 shades. Most digital imaging systems used in health care have dynamic ranges set at 10 bits (1024), 11 bits (2048) or in some advanced systems, 12 bits (4096).

What is a Pixel?

To the computer expert, a pixel has no particular size or shape—it is a *dimensionless point* in the image which has had a pixel value as signed to it. In some contexts, this may be the best way to think of a pixel. After all, an image can be enlarged by simply magnifying the size of these individual picture elements, while each element still displays its single pixel value across whatever area it occupies. Inside a CR reader, the pixels being sampled from an exposed PSP plate are circular in their initial shape, while the "final" pixels displayed from an LCD display monitor are squares. Each pixel is defined only by the pixel value it contains and by its relative location in the matrix, not by its size or shape.

However, in the imaging chain there are several different types of matrices involved in the acquisition of the image, processing of the image, and display of the image. In a DR system, the detector plate is made up of many hundreds of small detector elements that are essentially square in shape, and which have a fixed size since they are hardware devices. More importantly, the *hardware pixels* of an LCD (liquid-crystal display) monitor are made up of the intersections of flat, transparent wires crossing over each other to form an overall square shape. For the radiographer's purposes, it is probably best to visualize pixels as generally square in shape and having a set size. This will help to understand most of the concepts that are essential for clinical practice.

For radiography, we define a pixel as the *smallest element of the matrix or device that can rep-*

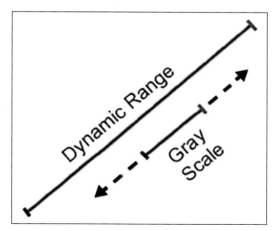

Figure 1-5. The dynamic range made available by the computer and its software must be long enough to allow windowing of the actual displayed gray scale up or down by several factors of 2.

r:sent *all pixel values within the system's dynamic range,* that is, all available pixel values.

Voxels, Dexels, and Pixels

Digital radiography (DR) and computed radiography (CR) both work on the basis of the *attenuation,* or "partial absorption," of x-rays as they pass through the human body. Each different type of tissue area within the body has an *attenuation coefficient* defined as the percentage or ratio of the original x-ray beam intensity that is absorbed by it. But, as illustrated in Figure 1-6, the x-ray beam is projecting data from 3-dimensional *volumes* of tissue onto a *2-dimensional image receptor surface.* At this surface, each square detector element (*dexel* or *del*) records *all* of the information within a square-shaped *tube* of tissue above it that extends all the way from the front of the patient to the back of the patient (for an AP projection). In effect, the attenuation coefficients for all of the tissues within this square-shaped tube are *averaged* together upon being recorded at the detector.

Because the x-ray beam is spreading out or *diverging* as it passes through the body, each of the squared tubes shown in Figure 1-6 actually expands somewhat as it passes from the front to the back of the body. Technically, then, these tubes are long *trapezoids* that are square in their cross-section.

We name these long square tubes *volume elements* or *voxels.* Each voxel represents the volume of tissue within the patient from which data will be collected and averaged by the *dexel* below it. The *averaged attenuation coefficient* determines the *pixel value* that will ultimately be assigned to the image pixel corresponding to this detector element. This pixel value must be selected from the dynamic range made available from the computer software, so the average attenuation coefficient that is measured by the dexel must be *digitized* or rounded to the nearest available value in the dynamic range.

As illustrated in Figure 1-6, we might summarize the radiographic process for a DR sys-

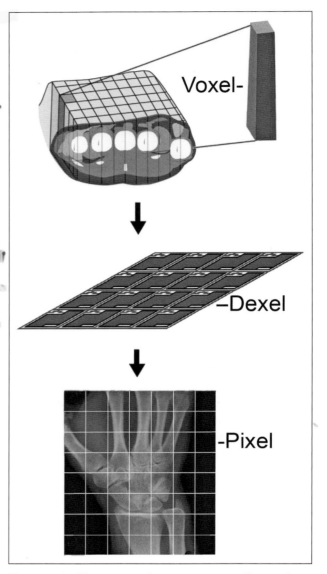

Figure 1-6. The averaged attenuation coefficient from each *voxel* within the patient is measured by the corresponding *dexel* of the DR detector plate below, then digitally processed to become an individual *pixel* in the displayed image.

tem as follows: To form a digital radiographic image, data from the *voxels* within the patient are collected by the *dexels* of the image receptor, then computer processed to become the *pixels* of the displayed image.

Chapter Review Questions

1. The first application of digital technology to radiographic imaging occurred in what year? 1979

2. "DR" systems convert the latent image carried by the remnant beam from x-rays directly into _____.

3. All digital images are laid out in pattern of pixel locations called a _____.

4. Unlike digital data, analog measurements are on a continuous scale that is infinitely _____.

5. To obtain discrete values for a set of digital data, analog measurements must be mathematically _____.

6. As long as the discrete units for a digital computer are smaller than a human can detect, digitizing data _____ read-out accuracy. (reduces or improves)

7. List the three steps to digitizing an image:

8. For a detector plate with hardware elements, collimation of the x-ray beam is analogous to which of these three steps (from question #7):

9. During the step of *quantization,* each pixel has a discrete pixel value assigned to it, selected from a predetermined scale of values called the _____ _____.

10. The full range of pixel values that any *hardware device* (such as a computer or a display monitor), can store is best labeled as its *bit* _____.

11. The range of pixel values made available by the combined hardware and software of an imaging system to build up the final displayed image is best labeled as the _____ _____.

12. The range of brightness levels visually apparent in the displayed image is best labeled as its _____ _____.

13. For radiography, we define a single *pixel* as the _____ element that can represent *all* the pixel values within the system's dynamic range.

14. To ultimately formulate a pixel for a DR or CR image, the first step is to measure and average the attenuation coefficients for all of the tissues within a square-shaped tube of tissue called a _____.

15. For a DR system, these measurements (from question #14) are detected and recorded by a _____ in the image receptor plate, then computer processed to become the pixel values of the displayed image.

1. 1979

2.

Chapter 2

CREATING THE LATENT IMAGE

■ ■

Objectives

Upon completion of this chapter, you should be able to:

1. List and give examples of the six general types of radiographic variables affecting the latent image carried by the remnant x-ray beam to the image receptor.
2. Define subject contrast and describe how it is controlled by different *ratios* of interactions and penetration through body tissues.
3. Quantify the effects of tissue thickness, physical tissue density, and tissue atomic number on the production of subject contrast in the remnant beam signal.
4. Describe the effect of increasing x-ray energy upon subject contrast in the latent image.
5. Explain why SID and mAs do *not* control subject contrast in the latent image.
6. State the primary role of radiographic technique in the digital age.
7. Describe the advantage or disadvantage the advent of digital imaging has for producing contrast and sharpness in the final displayed image.

The latent image will be defined throughout this book as the image information carried by the remnant x-ray beam and reaching the *image receptor, which is then passed from the image receptor into the computer processing system.* Three distinct types of variables determine the nature and quality of this information—technical variables, geometrical variables, and patient status.

Overview of Variables

Technical Variables

Technical variables include all the electrical factors set by the radiographer at the console—the kVp, the mA, and the exposure time or total mAs used to make the exposure, as well as the type of x-ray generator being used. Also included are the field size set by collimation, "cone" devices or masking materials, and the combination of inherent filtration within the x-ray tube and collimator and any compensating filters added by the radiographer.

Geometrical Variables

Geometrical variables include the size of the focal spot in the x-ray tube selected by the radiographer at the console, all of the various distances employed (the SID, SOD, and OID) while positioning the patient, the alignment between the x-ray beam, the anatomical part of interest, and the image receptor, and any angles placed on the central ray of the x-ray beam, the part, or the image receptor. All aspects of positioning the patient are geometrical in nature. Motion of the patient, the x-ray beam or the IR is also a geometrical factor.

A common misconception is that increasing the source-to-image receptor distance (SID) reduces exposure to the image receptor (IR) because of substantial absorption of x-rays by the air between the x-ray tube and the patient. As a *gas,* the density of air is less than *one-thousandth* the density of soft tissues in the body. Although there is some absorption of the x-ray beam by air molecules, these effects *never reach 5 percent of the simple geometrical effect of the inverse square law.*

It is essential for the student to understand that *nearly all* of the effect of distance changes on the intensity of radiation reaching the IR is due to the simple spreading out of x-rays as they travel away from their source, the x-ray tube. This spreading out is *isotropic,* that is, even in all directions, and can be predicted by examining the *area* across which the x-rays are distributed. For example, when the IR is moved twice as far away from the x-ray tube, the x-rays spread out over 2 squared or *4 times the area,* not just double the area, because they spread out twice as much lengthwise *and also twice as much crosswise* within the field. When the field is doubled in *both* length and width, the final result is *4 times the area* over which the x-rays are spread. This entire process has nothing to do with absorption by air molecules—it is strictly a geometrical phenomenon for anything that spreads out evenly.

The absorption of x-rays by air is less than 5% of the effect of the inverse square law. Because the absorption of x-rays by air is so minor, we do *not* generally take air absorption into consideration when making x-ray exposures. For our purposes, it is negligible. The geometrical effects of the inverse square law, however, are always important to consider.

Patient Status

Patient status encompasses body habitus and general body condition including age and health, as well as the impact of any specific diseases present upon the anatomy of interest, because diseases can alter both the density and the "average atomic number" of tissues, both of which affect their absorption of x-rays. Inter-

ventional changes such as surgical alterations, prostheses or other devices, casts, and the introduction of contrast agents into body cavities all affect x-ray absorption and the quality of the resulting latent image. Even the state of breathing, inhalation or exhalation, can change the latent image enough to affect diagnosis.

This is where the interactions of the x-ray beam with various atoms in the body results in a *signal—a latent image* being carried by the *remnant x-ray beam* to the IR behind the patient. Photoelectric interactions are primarily responsible for the production of *subject contrast* in the remnant x-ray beam between different areas of the beam from different tissues in the body part. Compton and Thompson interactions generate scatter radiation and are destructive to *subject contrast.* The details of these interactions may be found in radiography physics textbooks and will not be covered in detail in this textbook.

Attenuation is the partial absorption of x-rays by body tissues. It is essential to the production of an image signal in the remnant x-ray beam. If a particular tissue absorbed *all* x-rays striking it, the resulting image would be a *silhouette* of that organ presenting information only at its edges. Different body structures are fully demonstrated when each tissue allows *partial penetration* of the x-ray beam to a different degree than surrounding structures, such that each organ or body structure is demonstrated in the final image as a *shade of gray,* rather than "blank white" or "pitch black."

Creating Subject Contrast

X-Ray Interactions

Passing through the molecules of the human body, x-rays cause four types of interactions with atoms: The photoelectric interaction, Compton scattering, Thompson scattering, and characteristic interactions. It is not within the scope of this textbook to cover the physics behind these interactions, but we will review them as they pertain to image production and their implications for radiographic technique. Thompson interactions account for only 2 or 3 percent of all scattering interactions, so they are of little account. Given the small "size" of body atoms (having an average atomic number of only 7.6), the characteristic interaction results not in x-rays that could reach the image receptor (IR), but only in ultraviolet or visible light rays that cannot penetrate out of the body. Since they never reach the IR, these characteristic rays have no impact on the final image. Therefore, when it comes to radiographic technique, the two interactions of primary interest are the photoelectric effect and Compton scattering.

In the photoelectric effect, all of the energy of an incoming x-ray is absorbed by an inner electron shell of the atom, that is, the *entire x-ray photon is absorbed.* Since this leaves no exposure at the corresponding location on the IR, a miniscule "white" spot is left there, and *we can think of the photoelectric interaction as being primarily responsible for producing subject contrast in the latent radiographic image.* When photoelectric interactions are lost, poor image contrast results.

The Compton interaction is responsible for about 97% of all scatter radiation produced within the patient's body. Because scatter radiation is emitted in random directions, it lays down a "blanket" of exposure in the general area where it was produced, which obscures the visibility of the useful image just like fog can obscure the visibility of a bill board as seen from a highway. When this layer of fog is overlaid "on top of" all the details in the area, the result is reduced contrast in the latent image.

Finally, we should consider the very important x-rays which penetrate right through the patient's body *without undergoing any interaction* and reach the IR with all of their original energy. Each of these expose a miniscule "black" spot at the IR. The useful *signal* of the remnant x-ray beam is the "latent image" carried by it to the IR, and consists of varying degrees of penetrating x-rays mixed with photoelectric interactions that result in the various shades of gray representing different tissues. A structure such as a bone, that absorbs many x-rays by the photoelectric effect while a few penetrate clear through, will be depicted as a light gray density in the final image. An organ such as the heart,

absorbing much fewer x-rays by the photoelectric effect and allowing many more to penetrate right through, will be depicted as a dark gray structure in the final image. *The useful signal is created by variations in the photoelectric/ penetration ratio in different tissues of the body.*

Compton interactions, on the other hand, should be generally considered as *destructive* to the image. Because they are randomly distributed, *they carry no organized signal, but rather constitute a form of noise* which hinders the visibility of the useful image.

X-Ray Beam Energy

When the x-ray beam has higher overall energy, we call it a "hardened" beam which will better penetrate the body through to the IR. The "average" energy of the x-ray beam can be increased in two ways: By increasing the set kVp, and by adding filtration at the collimator. Filtration is not generally tampered with, and for our purposes here, can be considered as a fixed quantity, leaving kVp as the primary controlling factor for beam penetration.

As x-ray beam energy is increased and penetration improves, *all types of attenuating interactions within the patient's body decrease* (Figure 2-1). This overall ratio between penetration and interactions is what is mostly responsible for the creation of subject contrast. However, there is an additional effect caused by the ratio *between the different types of attenuating interactions that are occurring.*

See Figure 2-1 for example: At 40 keV, we find that about half of all interactions in soft tissue are photoelectric, and half are Compton, (the graph shows about equal amounts of both). Since we can think of half of all the interactions as "producing miniscule white spots," we expect soft tissue to appear as a medium-gray in the radiograph. But photoelectric interactions *plummet* as kVp is increased, with almost none occurring in soft tissue above 60 keV. At 80 keV, only penetrating x-rays and Compton scatter x-rays reach the IR under soft tissue areas, so they now appear so dark that most soft tissue organs are hard to distinguish.

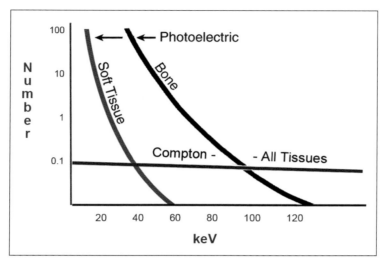

Figure 2-1. Within the patient, all attenuating interactions decrease at higher keV. Photoelectric interactions plummet quickly, while Compton interactions only decrease slightly, leaving Compton as the dominant interaction at higher keV's.

Now, let's consider *bone* tissue: Examining Figure 2-1 again, at 40 keV, in bone there are *one hundred times* more photoelectric interactions occurring than Compton interactions, (10/0.1). When the keV is increased to 80, both photoelectric and Compton interactions decrease, but Compton only decreases very slightly (less than 2%), whereas photoelectric interactions *drop very sharply* to about 1/30 the prior amount (10/0.3). Yet, we find that more than one-half of the interactions in bone are still photoelectric. Thus, at 80 keV we still expect bone to appear as a light-to medium-gray density. An 80-kVp radiograph demonstrates bone structures very well, but soft tissue organs very poorly. This is why contrast agents must be used to make the stomach or bowel show up.

Tissue Thickness

As a body part becomes thicker, all types of attenuating interactions are affected by equal proportions, (photoelectric and Compton interactions go down together). The general attenuation of the x-rays is exponential and follows a practical rule of thumb: *For every 4-5 centimeters increased part thickness, attenuation is doubled, so radiographic technique must be doubled to compensate.*

Tissue Density

The physical density of body tissues (the *concentration* of their mass) is *directly proportional* to the number of all attenuation interactions taking place. If an organ, such as the liver, is twice as dense as the muscles surrounding it, it will absorb twice as many x-rays and show up as a lighter gray area on the final radiograph. The different types of interactions are all affected by equal proportions, (photoelectric and Compton interactions go down together).

Tissue Atomic Number

A high atomic number for an atom means not only that it has a "large" nucleus, but many electrons. A peculiar aspect of atoms is that *they all have a roughly equal diameter*, occupying about the same amount of space, no matter how many electrons are "packed" into them. This means that for atoms with high atomic number, the many electrons are much more *concentrated*, something physicists refer to as an atom's *electron density*.

With many more electrons *per cubic angstrom*, x-rays are much more likely to strike and interact with them—We expect x-ray absorption to go up exponentially for "larger" atoms with more electrons, and this is the case. In fact, x-ray absorption increases by the *cube* of the atomic number. For example, let's compare the two elements *hydrogen* and *carbon,* both of which are plentiful within the human body: The atomic number for hydrogen is 1, and 1 cubed (1 x 1 x 1) is 1. The atomic number for carbon is 6, and 6 cubed (6 x 6 x 6) is 216. This means that carbon is *216 times* better at absorbing x-rays than hydrogen.

In practical radiography, this atomic number effect is the main reason that *positive contrast agents* and *bones* absorb x-rays so much more than soft tissues.

Contrast Agents

The atomic number of iodine is 53, and that of barium is 56. Compared to an *effective* (average) atomic number for soft tissue of 7.4, these are approximately 7 times higher atomic numbers. Using the cube rule just described in the last section, we find that these *positive* contrast agents are about 343 times more effective at absorbing x-rays than soft tissue.

Gasses such as air are sometimes also used as *negative* contrast agents. In this case, the primary change in absorption is due to the extreme difference in physical *density,* as described in the above section on *tissue density.*

Summary of Variables Creating Subject Contrast

Taken together, we find that the lungs show up against the soft tissues of the heart and surrounding musculature *primarily because of the difference in physical density.* The lungs are insufflated with air, which is a gas. Gasses generally have 1/1000th the physical density of soft tissue, and even though density is proportional to x-ray absorption, this is such a huge difference that the lungs are indeed demonstrated against the soft tissues around them. This is true even though the *effective* (average molecular) atomic number of air is 7.6 and that of soft tissue is 7.4, nearly equal.

Bones show up dramatically on radiographs *primarily because of the difference in atomic number.* Bone has an effective (average) atomic number of roughly 20. When we cube this number and that for soft tissue (7.4), we get 8000 and 405, respectively. Taking the ratio 8000/400, we see that bone absorbs about 20 times more x-rays than soft tissue on account of its atomic number. However, bone is *also twice as dense* as soft tissue, and absorbs twice as many x-rays on account of this density difference, for a total absorption of about 40 times that of soft tissue.

Table 2-1 is a comprehensive list of the variables that control subject contrast in the latent image carried by the remnant x-ray beam to the image receptor.

Why mAs and SID Do Not Control Subject Contrast in the Latent Image

Milliampere-seconds (mAs) is directly proportional to the *intensity* of the primary x-ray beam emitted from the x-ray tube. When the mAs is doubled, twice as many electrons strike the anode, and precisely twice as many x-rays are produced *at every energy level.* That is, there

Table 2-1
SUBJECT CONTRAST: VARIABLES

X-ray Energy (kVp, filtration) Tissue Thickness Tissue Density Tissue Atomic Number Presence of Contrast Agents

are twice as many 30-kV x-rays, twice as many 45-kV x-rays, and twice as many 60 kV x-rays, all the way up to the set kVp or *peak energy*.

Within the patient, twice as many x-rays penetrate through to the IR without undergoing any interactions, twice as many photoelectric interactions occur, and twice as many Compton scattering events occur. There is no reason for the *proportions* between these interactions to change, because there has been no change made to the relative energy levels of the x-rays or the penetration characteristics of the x-ray beam. Everything has simply been doubled.

At the image receptor, we might think of the effect as twice as many miniscule "white spots" being created, but also twice as many "fogged" spots created, along with twice as many "black" spots created. The overall exposure has been doubled, so the latent image *would* appear twice as dark overall, yet the *subject contrast* remains unchanged.

It is important to note that the overall density of an image can be doubled without changing the contrast. For example, let's begin with two adjacent densities that measure 1 and 2 respectively, then double the overall exposure by doubling the mAs. The density that originally measured 1 will now measure 2, and the density of 2 will now measure 4. Contrast is the *ratio* of difference between the two adjacent densities. The original contrast was $2/1 = 2$. The new contrast is $4/2 = 2$. In other words, even though the overall image is twice as dark, the second density is *still twice as dark* as the first density. Contrast has not changed.

(Now, an *extreme* increase in mAs, such as 8 times the mAs, leading to gross overexposure, can continue to darken lighter areas while those that are pitch black cannot get visibly any darker. At this extreme, image contrast can be destroyed. However, such an extreme increase in technique is not something that one encounters in daily radiographic practice—it is the exception rather than the rule, and so should not be used to generalize the effects of mAs. The above example, where mAs was doubled, is a realistic scenario in practice, and it can be clinically demonstrated that it does not alter the image contrast.)

Changes made to the source-to-image distance (SID) follow precisely the same logic, even if the physics is somewhat different. SID simply alters the *concentration* of x-rays according to the inverse square law, as they spread out more at longer distances, and spread out less at short distances. So far as the image receptor (IR) is concerned, the result in exposure is no different than changing the mAs. In other words, whether the x-rays are twice as concentrated because the x-ray tube was moved closer, or whether twice as many were produced in the first place, the result is the same at the IR—twice the exposure over a given area.

Ultimately, at the IR, both mAs and SID control the intensity of exposure, but not the *distribution* or *ratio* of different types of interactions. Therefore, neither mAs nor SID "control" subject contrast in the Latent Image. This actually makes life less complicated for the radiographer, because we can focus on using kVp to control penetration and subject contrast without worrying about these other factors.

Role of Radiographic Technique in the Digital Age

As will be fully described in Chapter 4, the primary role for the set radiographic technique in the digital age is to simply provide sufficient signal reaching the IR, that is, enough information for the computer to process. This depends *both* on beam intensity (controlled by the mAs and the SID) *and* on beam penetration (controlled by filtration and the kVp). With digital equipment, there is great leeway for using higher kVp levels to achieve this goal, with the added benefit of reduced patient dose when the mAs is compensated according to the "15 percent rule."

Subject Contrast and Sharpness

It is important to distinguish between the qualities of the *latent image* carried by the remnant x-ray beam to the image receptor, discussed here, and the qualities of the final displayed image. The remnant x-ray beam carries the *signal*—the information organized by the passage of the x-ray beam through the various body structures of the patient. This signal possesses *subject contrast* between different areas of x-ray intensity. It includes noise in the form of scatter radiation and quantum mottle. It includes an inherent spatial resolution determined by the geometry of the x-ray beam in relation to the patient. Because of this same beam-patient geometry, the latent image is also magnified to some degree and may be distorted.

Although the image will be radically altered in nearly every way by digital processing before it is displayed, these qualities of the remnant beam signal at the IR can set important *limitations* on what digital processing can accomplish.

First, within the dynamic range, there must be a certain minimum degree of distinction between one shade of gray and the next, that is, a minimum level of *subject contrast* present in the data set that is fed into the computer. This is theoretically true, yet the incredible ability of digital processing to enhance subject contrast makes it a minor point in practice. Compared to conventional film imaging, digital equipment has more than *10 times* the capability to demonstrate tissues with low initial subject contrast. In general, with digital technology, we need not concern ourselves with contrast resolution in daily practice, and it is just this flexibility which allows us to use higher kVp levels, reduce or eliminate grid use, and make other practical changes with an eye toward minimizing patient dose.

Sharpness, on the other hand, is of more concern. Digital equipment has less capability for spatial resolution than the older film technology. A key issue, for example, is the interplay between focal spot size and the inherent resolution of digital display systems. A typical LCD monitor has hardware pixels of 0.1 or 0.2 mm in size. The small focal spot in the x-ray tube is typically 0.5 to 0.6 mm in size. The large FS is 1.0 to 1.2 mm, much larger, but because of the extremely high SOD/OID ratio in the geometry of the beam itself, the resulting "effective pixel size" at the surface of the IR is reduced to a level comparable to the pixels of the display monitor.

The important point here is that these are close enough that *poor practice in making the original x-ray exposure can, in some cases, reduce the sharpness below what the digital system is capable of displaying.* For one thing, this means that radiographers must still be conscientious to *use the small focal spot whenever feasible,* such as for smaller extremities and for pediatrics. Beam geometry, including *maximum feasible SID,* and *minimum OID* continues to be of prime importance. Optimum focal spot, beam geometry, and positioning must continue to be used in daily practice to maximize the initial sharpness of the image being sent into the computer.

Variables Affecting the Quality of the Final Displayed Radiographic Image

By way of an overview, below is a list of the many types of variables that digital image capture, digital processing, and electronic image display introduce into the quality of the final image that is displayed on a monitor. These will all be fully discussed in the following chapters.

1. Characteristics of the IR, discussed in Chapter 10
2. Digital Processing of the image, discussed in Chapters 5 through 9
3. Characteristics of the Display Monitor, discussed in Chapter 11
4. Ambient Viewing Conditions, discussed in Chapters 11 and 12

Chapter Review Questions

1. What are the three general types of variables that determine the qualities of the latent image carried by the remnant x-ray beam to the image receptor?
2. Nearly all of the effect of distance (SID) changes on the intensity of x-rays reaching the image receptor is due to the simple _____ out of x-rays as the travel.
3. The partial absorption of x-rays by body tissues is called _____.

4. Which type of x-ray interaction within the patient's body is primarily responsible for the creation of *subject contrast* in the signal of the remnant x-ray beam?

5. The useful signal is created by variations in the photoelectric-to-_____ ratio in different tissues of the body.

6. As x-ray beam penetration improves, _____ types of attenuating x-ray interactions within the body decrease.

7. As x-ray beam energy is increased, less subject contrast is produced in the remnant beam because the occurrence of _____ interactions *plummets* compared to other interactions.

8. As a body part becomes thicker, all types of attenuating interactions are affected by equal _____.

9. Radiographic technique must be doubled for every _____ centimeter increase in part thickness.

10. The physical density of a tissue is _____ proportional to the number of attenuating x-ray interactions that will occur in it.

11. X-ray absorption increases by the _____ of the atomic number of the atoms in tissue.

12. Against a background of soft tissue, positive contrast agents such as iodine or barium show up on a radiograph primarily because of their difference in _____.

13. Against a background of soft tissue, gasses such as air show up on a radiograph primarily because of their extreme difference in physical _____.

14. SID and mAs are NOT considered as controlling factors for subject contrast because, although they affect the overall number of interactions occurring, they do NOT change the _____ between different types of interactions.

15. The primary role for radiographic technique in the digital age is to simply provide _____ signal reaching the image receptor.

16. Sufficient signal at the image receptor depends upon both the _____ and the _____ of the x-ray beam.

17. Which is the greater concern for digital units in daily practice, producing good image contrast or producing a sharp image?

18. Compared to the older film technology, digital equipment has _____ times the capability for demonstrating tissues with very low subject contrast.

19. Even though the pixels on a display monitor are fairly large (0.1 to 0.2 mm), it is possible for improper use of the large focal spot during the initial exposure to reduce image _____ below what the digital processing and display systems are capable of.

Chapter 3

QUALITIES OF THE DIGITAL RADIOGRAPH

Objectives

Upon completion of this chapter, you should be able to:

1. List and define the six qualities of the gross, anatomical image used for medical diagnosis.
2. Relate the visibility qualities to window level, window width, and signal-to-noise ratio.
3. State the "ideal" levels of each image quality for clinical diagnosis.
4. Describe how image contrast is properly measured and analyzed.
5. List eight general categories of image noise.
6. Distinguish display magnification from geometric magnification for the image.
7. Distinguish qualities of the latent image from those of the displayed image.
8. Define resolution, contrast resolution, and spatial resolution as physicists use the terms.
9. Relate exposure trace diagrams to contrast resolution and spatial resolution.

Before we can have any clear discussion about the effects of either radiographic technique or digital processing on the image in subsequent chapters, we must carefully distinguish between the actual qualities of the final, electronically displayed (TV) image and the "latent image" created in the image receptor by the x-ray beam. It will also be helpful to point out how these differ from the qualities of a conventional radiograph recorded on film.

Qualities of the Final Displayed Digital Image

Let's begin by making a comprehensive, but concise list of what those qualities are, presented as a "hierarchy" of image qualities in Figure 3-1. Definitions for each of these terms are presented in the Glossary.

Brightness

Brightness is intuitive to understand; It refers to the intensity of light for the overall image or any portion of the image. All pixel brightness levels within the anatomy should be neither completely white nor pitch black, but an intermediate shade of gray ranging from very dark gray along a broad scale to very light gray. To achieve this, at times the displayed image will need to be *levelled*. This is an adjustment to the *window level* discussed under postprocessing in Chapter 7. For most radiographic equipment, the higher the window level number, the darker the image.

Density is the darkness of any portion of the image, so it is the opposite of brightness. Although this term is a hold-over from the days of film radiography, it is still often used and is appropriate when discussing certain radiographic variables, especially in regard to a printed image.

20

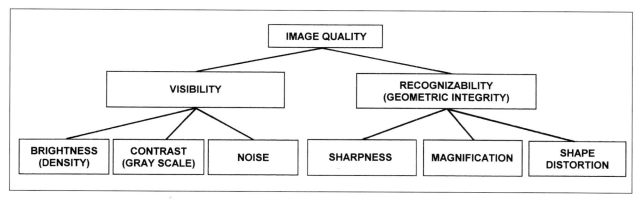

Figure 3-1. Hierarchy of radiography image qualities. (From Q. B. Carroll, *Radiography in the Digital Age,* 3rd ed. Springfield, IL: Charles C Thomas, Publisher, Ltd., 2018. Reprinted by permission.)

The ideal amount of brightness for an image is an intermediate level that is considered optimum for the human eye (*neither maximum nor minimum*).

Contrast and Gray Scale

Contrast is the *percentage or ratio* between two adjacent brightness or density levels, Figure 3-2. A certain minimum level of contrast is necessary between two adjacent details for the human eye to distinguish them apart. Contrast can be calculated by dividing measurements taken from a photometer or densitometer. Generally, a high-contrast image is easier to see than a low-contrast one. Contrast is essential to the *visibility* of any image.

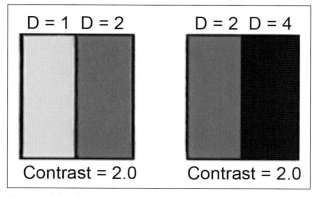

Figure 3-2. Contrast is the *ratio* between two adjacent densities. These two examples both have a contrast of 2.0 (2/1 and 4/2) even though the image on the right is darker overall.

Gray Scale is the range or number of different brightness levels or densities present in the image, as counted on a scale from white to black. The significance of gray scale is that when more different shades of gray are available to be recorded in the image, more different types of tissues and anatomical structures within the body can be represented. Increased gray scale is associated with *more information* present in the image, Figure 3-3. In the displayed image, gray scale is adjusted by *windowing.* This is an adjustment to the *window width* discussed under postprocessing in Chapter 7. For most radiographic equipment, the higher the window width number, the longer the gray scale.

For the human eye, blank white and pitch black *fix* in place the extreme ends of the scale of perceived densities. (That is, you can't get any lighter than blank white, or any darker than pitch black.) The result is that, as the gray scale is increased between these extremes, the difference *between* densities or brightness levels must decrease. For example, as shown in Figure 3-4, if you fix the ends of a staircase between the ground floor and the second floor of a home, and the ground floor ceiling is 10 feet high, you can have 5 steps on the staircase that are each 2 feet apart in height, or you can increase the number of steps to 10 only by reducing the difference between them to one foot each, but you cannot have 10 steps and keep them 2 feet apart. As the number of steps increases, the dif-

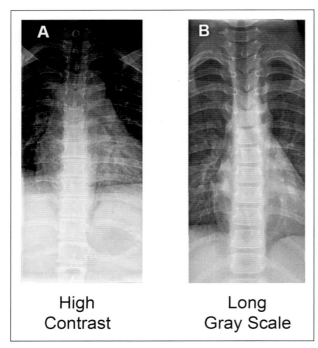

Figure 3-3. Although high contrast (*left*) can increase the visibility of particular details, long gray scale (*right*) demonstrates *more* different types of tissues and provides more diagnostic information.

ference between them must decrease. For this reason, gray scale is generally considered as *opposite* to contrast. (Increasing the *window width* reduces contrast.)

We have stated that contrast is essential for the visibility of details in the image, but that increased gray scale provides more information. Which is more important, the quantity of information present or the ability to see it well? The answer is that both are equally important. Therefore, the ideal amount of contrast or gray scale in an image is an intermediate level that is *optimal (not maximum nor minimum).*

Measuring Contrast

When actually *measuring* contrast in a radiographic image, it is essential to compare two densities that both *represent tissues within the anatomy,* one darker tissue and one lighter tissue. On an AP knee view, for example, the darker soft tissue of the joint space can be compared to a homogeneous area within the hard, cancellous outer bone. *Never use the "raw exposure" background density for the darker of the two density (brightness) measurements.* Since the background density is *always* pitch black and cannot be darkened any further, using it for contrast measurements leads to false conclusions about the image. A classic example is the notion that a "darker image always has less contrast"—we have just proven this false in Figure 3-2. There is no direct causative relationship between overall image darkness and the contrast between organs such as the

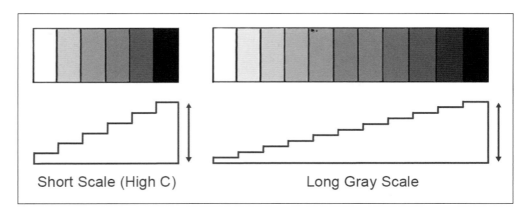

Figure 3-4. When "stretched" between the extremes of blank white and pitch black, *gray scale* is like a staircase reaching a fixed total height (*arrows*): The more steps there are, the less the distance or difference between one step and the next.

heart against the lung field or a bone against a soft tissue background. It is the contrast between anatomical tissues we are vitally interested in.

(On larger body parts such as a large abdomen, there may not even *be* any "raw" background density to measure, so the same type of experiment will yield different results compared to a knee image in which the background density is used.) Contrast must always be measured *within* the anatomy.

It is also hazardous to use *subtraction* rather than a *ratio* or *percentage* to calculate the contrast between two density measurements. When we describe a density as being "twice as dark" as another, we are intuitively using a ratio (2 to 1), *not* a subtracted difference. Let's illustrate with a simple example: Suppose we have two density (or brightness) measurements of 1.0 and 2.0. We then add a "fog" density across the whole image with a value of 1.0. Adding this 1.0 to the two original numbers, we now have 2.0 and 3.0. Using *ratios*, the contrast of the original image is $2/1 = 2$, and the contrast of the new fogged image is $3/2 = 1.5$. Note that the contrast has gone down from 2 to 1.5 due to fog. This is the correct result for a fogged image.

If we now perform the same calculation using *subtraction*, the contrast of the original image is $2 - 1 = 1$, and that of the new fogged image is $3 - 2 = 1$. Apparently, contrast was unaffected by fog. This is, of course, incorrect. Perform the same exercise for mAs or SID, and we find that using subtraction, these variables change contrast, also false. Repeat these calculations using *ratios* for mAs and SID, and we find that they do not control contrast, the correct relationship, because these two variables change all image densities by the same proportions. For example, doubling the mAs doubles *all* densities, so the second image in our example would have the densities 2.0 and 4.0, respectively, and the contrast is still 2.0.

Noise and SNR

Noise consists of any form of non-useful contribution to the image which interferes with the *visibility* of useful information (anatomy or pathology of interest. Three examples are demonstrated in Figure 3-5. Noise is always destructive, so the "ideal" amount for image quality is generally the minimum *noise possible*. (For quantum

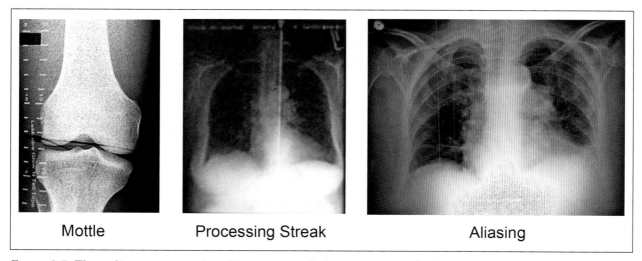

Mottle   Processing Streak   Aliasing

Figure 3-5. Three disparate examples of image noise: Left, quantum mottle, the most common form of noise for digital radiography; Middle, a streak artifact from a CR reader (Courtesy, State University of New York. Reprinted by permission.); Right, aliasing or *Moire effect* caused during CR processing. (From Q. B. Carroll, *Radiography in the Digital Age,* 3rd ed. Springfield, IL: Charles C Thomas, Publisher, Ltd., 2018.)

Table 3-1
TYPES OF IMAGE NOISE

1. Scatter radiation
2. Off-focus radiation
3. Quantum mottle
4. Material mottle
    A. CR PSP plate crystal layer defects
    B. DR or DF indirect-conversion crystal layer defects
    C. Image intensifier input phosphor crystal layer defects
    D. Fiber-optic bundle defects
5. Aliasing artifacts
    A. Insufficient digital sampling
    B. Magnification (zoom) of displayed image
    C. Grid line frequency vs. sampling frequency
6. Electronic noise
    A. Direct-conversion DR active matrix arrays (AMAs)
    B. CR reader hardware
    C. Image intensifier
    D. Charge-coupled devices (CCDs) or CMOS
    E. Computer hardware, including ADCs and DACs
    F. Display systems
7. Algorithmic noise (software defects)
8. Exposure artifacts
    A. Extraneous objects
    B. Grid lines
    C. Superimposition of anatomical structures within patient
    D. False images created by movement or beam geometry

mottle, we accept the *minimum feasible,* taking patient dose into account.)

Table 3-1 presents a list of the various types of noise that can be present in any electronically displayed image. These are many. Some authors have restricted the concept of noise to quantum mottle, but this is misleading: In addition to the x-ray beam, mottle can also originate from materials used in the image receptor, from small variations in electrical current, and even from eccentricities in computer algorithms. Scatter radiation is also destructive to the contrast between small details. Grid lines and other artifacts interfere with the visibility of details by *superimposing* them with a blank or nearly-blank white image, effectively "blocking them out."

A related image quality concept that is important for radiographers to understand is *signal-to-noise ratio (SNR).* The *signal* is defined as the or-

ganized, useful information reaching the image receptor (IR). It is carried to IR by the remnant x-ray beam in the form of different areas of x-ray intensity which possess *subject contrast* between them. Signal-to-noise ratio, then, is the *ratio of useful, constructive information to non-useful, destructive input* forming the latent image. Clearly, the ideal SNR for any image is the *maximum SNR.* This is achieved by both proper radiographic technique and by good positioning practices upon exposure.

Sharpness (Spatial Resolution)

Sharpness (sometimes called *spatial resolution*) is defined as the abruptness with which the edge of a structure "stops" as one scans across an area of the image (or how "quickly" the density changes from the structure to the background density). Imagine driving a microscopic sports car across the image at a constant speed: When passing over the edge of a bone image into the darker background density of soft tissues, if the density suddenly changes from white to dark, the edge of the bone image is sharp, if the density gradually fades from white to dark, the edge of the bone image is blurry and the spatial resolution is poor. Figure 3-6 demonstrates this concept.

Sharpness is affected by a number of variables: By the focal spot, beam projection geometry and any motion during exposure, by digital processing, and especially by the quality of the display monitor and any changes to magnification of the displayed image by the radiographer. *Maximum* sharpness is always the goal.

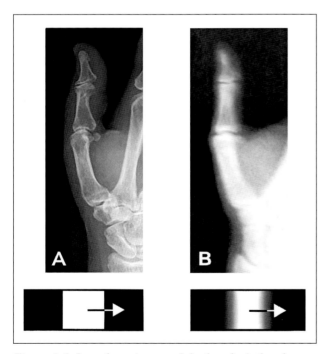

Figure 3-6. In a sharp image of the hand, *A*, the change from light to dark at the edge of a detail is *abrupt* or *sudden* as one scans across the image, (left, bottom). For a blurry image of the hand, *B*, the edge of a bone *gradually* fades from light to the dark background (*right, bottom*). (From Q. B. Carroll, *Radiography in the Digital Age,* 3rd ed. Springfield, IL: Charles C Thomas, Publisher, Ltd., 2018. Reprinted by permission.)

Shape Distortion

Shape Distortion is defined as any difference between the shape of a structure in the displayed image and the actual shape of the real structure itself within the patient's body. It is primarily controlled by positioning and the alignment (centering and angling) of the x-ray beam during exposure. Distortion is not generally altered by digital processing or display. *Minimum* distortion is always the ideal.

Geometric Magnification

Geometric Magnification is the difference between the size of a structure in the displayed image and the actual size of the real structure itself, created by the geometry of the original x-ray projection. While special magnification techniques can sometimes be used to make a particular structure more visible in the image, the general rule is that the image should be representative of the actual size of the real structures, (especially the heart for chest radiographs), and that geometrical magnification should be kept at a *minimum*.

Display Magnification

Display magnification is magnification created by applying "zoom" or "magnify" features of the display monitor, or by any form of post-collimation which changes the field of view displayed within the physical area of the monitor screen. Some magnification can assist in visualizing small structures. However, excessive magnification can result in a "pixelly" image, where individual pixels become apparent and the sharpness of image details is lost as shown in Figure 3-7.

At certain levels of magnification, a moire or aliasing pattern can also appear in the displayed image which is quite distracting (see Figure 9-11 in Chapter 9). Thus, display magnification can be diagnostically useful if not applied to excess.

As shown on the left in Figure 3-1, brightness, contrast, and noise combine to determine the general *visibility* of details in the image. On the

Figure 3-7. Example of a *pixelated* image (right). Excessive zoom or magnification of the image on the display monitor eventually leads to pixilation, loss of sharpness and aliasing artifacts.

right, sharpness combines with shape distortion and magnification to determine the geometrical integrity of the details in an image. Whereas the three qualities on the left half of the diagram in Figure 3-1 determine whether details in the image are *visible,* the three geometrical qualities on the right determine whether those details can be *recognized* for what they are. High visibility combines with maximum *recognizability* to determine the overall quality of the image.

Qualities of the Latent Image Captured at the Image Receptor

It is important not to confuse the qualities of the final displayed image with those of the *latent image* carried by the remnant x-ray beam to the image receptor. The remnant x-ray beam carries the *signal*—the information organized by the passage of the x-ray beam through the various body structures of the patient. This signal possesses *subject contrast* between different areas of x-ray intensity, as shown in Figure 3-8. It includes noise in the form of scatter radiation and quantum mottle. It includes an inherent spatial resolution determined by the geometry of the x-ray beam in relation to the patient. Because of this same beam-patient geometry, the latent image is also magnified to some degree and may be distorted.

With the single exception of shape distortion, every one of these aspects of the radiographic image will be modified by digital processing before the final image is displayed on an electronic monitor. Table 10-3 in Chapter 10 lists those aspects of digital processing which now "control" the various qualities of the final displayed digital image, compared with the old "controlling factors" of traditional film radiography. With conventional film radiography, the radiographic technique set at the x-ray machine console controlled all of the visibility characteristics—the brightness, contrast and noise—of the final image. It no longer does. As described in Chapters 6 and 7, digital rescaling and window levelling now "control" the final image brightness, gradation processing (LUTs) and the window width control adjust the final image contrast, display monitor characteristics and displayed field of view are the primary determinants of the final image sharpness and magnification. Then, what is the function of radiographic technique with digital equipment? This is the subject of the next chapter.

The crucial point we are trying to make here is to not confuse these final displayed image qualities with those of the latent image carried by the remnant x-ray beam. Those exposure qualities are controlled by the set radiographic technique, positioning, and patient condition. But, they describe only the image that is captured by the image receptor and *fed into the computer system,* not the image coming out of the computer and sent to display. As stated by the American Association of Physicists in Medicine

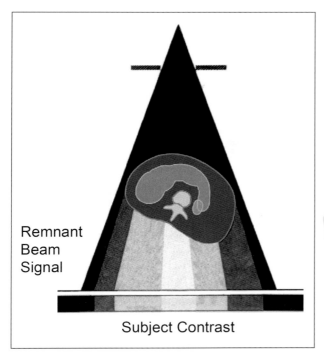

Remnant Beam Signal

Subject Contrast

Figure 3-8. The remnant x-ray beam carries subject contrast and other qualities of the *latent image* to the receptor below. However, digital processing will alter all but one of these qualities (shape distortion) before the final image is displayed.

(Task Group 116, *Medical Physics,* Vol. 36, No. 7, July 2009), "Unlike screen-film imaging, image display in digital radiography is independent of image acquisition." In other words, the qualities of the final displayed image have been "decoupled from" the conditions of the original x-ray exposure. A common mistake is to continue to treat x-ray exposure factors as "primary controlling factors" over the qualities of the final image, when in fact they are now relatively minor "contributing factors."

Resolution at the Microscopic Level

Confusion may result when physicist terms are mixed or conflated with the terms radiographers are used to in describing image qualities. An example is the adoption by the American Registry of Radiologic Technologists in 2017 of the term *spatial resolution* to describe *sharpness* in the radiographic image. Physicists examine the image *at a microscopic level,* and as we shall see, this makes a great deal of difference, because radiographers are concerned with the readily apparent qualities of the gross image in daily practice. In effect, the physicist is concerned with the resolution of a single *dot* or *detail* in the image, while the radiographer is concerned with the true representation of a *bone* or other anatomical structure.

Resolution is specifically defined as *the ability to distinguish two adjacent details as being separate from each other.* When one examines a single detail or dot in the image, all image qualities are reduced to just two characteristics. Physicists refer to these as *contrast resolution* and *spatial resolution.* Now, take a look at Figure 3-9 and note that two, and only two, distinctions need to be made in order to identify one dot as being separate from other surrounding dots: First, the dot itself must be distinguishable against the *background* density or the field behind it. It must be brighter or darker than its surroundings, otherwise it cannot be made out at all as a detail. Physicists call this quality *contrast resolution.* Second, there must be enough of the background field *between* this dot and any other nearby dot to visually *separate them in space.* Physicists refer to this quality as *spatial resolution.*

Figure 3-9 helps clarify that for the physicist, the only question is, "Can the dot be made out (overall resolution) against the background (contrast resolution) and separate from other dots (spatial resolution)?" The next section on exposure trace diagrams helps the student to visualize these concepts.

Now compare Figure 3-9 with Figure 3-1 at the beginning of this chapter, and it will quickly become apparent that contrast resolution best correlates with *visibility* of the image, and spatial resolution best correlates with *geometrical integrity* or *recognizability* of the image. That is, the two related questions for the radiographer are, "Can you see the anatomical structure?" and, "Can you recognize the anatomical structure?" Note that *none of the six image qualities listed along the bottom line of Figure 3-1 can, or should be, translated as "contrast resolution" or "spatial resolution."* Nor are these six characteristics of any particular interest to the physicist. Why not?

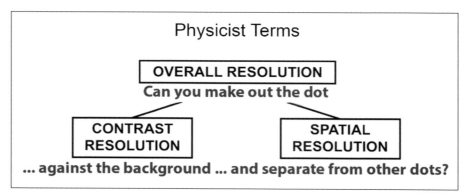

Figure 3-9. Physicists evaluate images only on the basis of their microscopic qualities, the *contrast resolution* and *spatial resolution* of a small dot or line. For the radiographer, these correlate to *overall visibility* and *overall recognizability*, and should not be confused with the six clinical qualities of a gross image such as a bone (see Figure 3-1, *bottom*).

The answer is *scale*. Scale literally determines our physical reality. And, scientific principles apply only on certain scales. For example, cosmological physics deals with the realm of the incredibly large, such as the universe itself. At this huge scale, Einstein's theory of general relativity applies, and, for example, time runs much slower near very massive objects such as stars. But, here on earth, we need not worry about the effects of a very large building slowing our time down because the scale of our daily human experience is too small to "feel" the effects of relativity. Likewise, quantum physics studies the realm of unimaginably small things such as an electron, which can turn back and forth between being a wave and being a particle, and can effectively disappear and reappear in different locations spontaneously. In our daily experience, we needn't worry about our car disappearing like an electron, because we are on a completely different scale of reality.

The six image qualities listed at the bottom of Figure 3-1 apply *only* to gross images such as the image of a bone. Physicists study the *microscopic* image, where these qualities become irrelevant. For example, the size and shape of a single dot in the image are determined by the size and shape of the dexel that detected the information and the pixel that displays it. Therefore, it is irrelevant to ask if a single dot is "distorted" or "magnified"—all dots in the image are the same size and shape. Likewise, it is meaningless to ask

if a single dot has "noise" impeding its visibility—noise in the image consists of dots just like useful data does.

For the radiographer, the image of a *bone* may be noisy (such as covered with mottle), the bone may be magnified, and its shape may be distorted by poor positioning or an improper beam angle. For the physicist, none of these questions would apply to a single dot. Why, then, should the radiographer need to know anything about *contrast resolution* and *spatial resolution*? Because these are physics standards that can be used to compare different types of radiographic equipment. When deciding which x-ray machine to buy, it is useful to be able to take a simple resolution device, such as the line-pair resolution template shown in Figure 3-10 and compare which machine produces the best resolution, without having to place a patient in the x-ray beam.

Exposure Trace Diagrams

A visual device called the exposure trace diagram can help understand the concepts of contrast resolution and spatial resolution. This type of diagram was invented during the days of film radiography when more exposed areas of the developed film literally had thicker deposits of blackened silver on them, while lighter areas such as bone images were created by stripping off silver during chemical development, such that only a thin layer was left on the film. In Figure

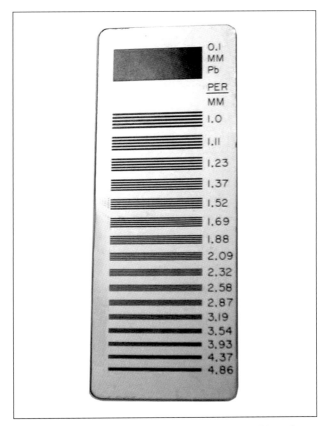

Figure 3-10. The resolution test template used by physicists consists of alternating slits and strips of lead foil to form ever smaller "line pairs." Exposure to x-rays results in a measurement of *modulation transfer function* (see Figure 3-12) or *spatial frequency* in line-pairs per millimeter (LP/mm).

3-11 we see a typical exposure trace for a single light spot in the image. The vertical thickness of the shaded area represents the amount of x-ray exposure received at the image receptor. In the middle of the image, we can consider the vertical depth of the "pit" as the contrast for the light spot or detail.

Interestingly, we can also represent the *sharpness* of this image, because the edges of the "pit" do not drop straight downward, but follow a *slope* that represents the amount of *penumbra* or *blurriness* of these edges. The *wider* these slopes are side-to-side, the more blurred the edges of the image. Now, comparing this diagram with the physicists' terms listed in Figure 3-9, we recognize that *contrast resolution* corresponds to the *vertical dimension* or depth of the exposure trace, and that *spatial resolution* relates inversely to the *horizontal dimension* of the edges in the exposure trace. That is, a *steeper* slope has less horizontal spread, indicating less penumbra or blur, and this translates as higher spatial resolution.

When the resolution template in Figure 3-10 is exposed to x-rays, the resulting image is a series of lines, each of which could be considered in cross-section as an exposure trace diagram. The series of up-and-down modulations appears as a sign-wave (Figure 3-12). This sign-wave is referred to by physicists as the *modulation transfer function (MTF)*, and it measures how

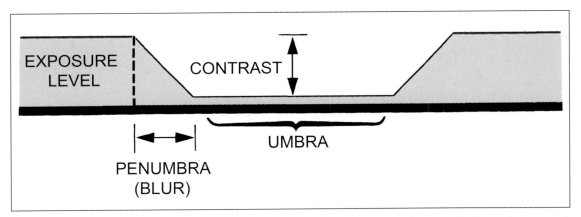

Figure 3-11. An exposure trace diagram illustrates the two basic characteristics for a single dot or line in an image: The vertical dimension of the "pit" corresponds to *contrast resolution* for the physicist. The horizontal length of a slope represents the amount of blur or penumbra and corresponds inversely to *spatial resolution* for the physicist.

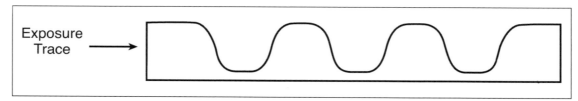

Figure 3-12. The *modulation transfer function (MTF)* sine-wave is, in effect, a series of exposure trace diagrams (Figure 3-11) created by exposing a line-pair resolution template (Figure 3-10).

well the difference between adjacent small details is transferred from the real structures within the patient to the image receptor.

With a little imagination, you can visualize the waves in Figure 3-12 sliding horizontally closer and closer together as the lead foil lines in the test template (Figure 3-10) become smaller and smaller. As shown in Figure 3-13, eventually the crests of the waves begin to *overlap,* and the vertical depth of the troughs is lost. This indicates that two details have become so close together that *their penumbras begin to overlap.* The loss in vertical depth (Figure 3-13*C*) indicates that contrast resolution is now being affected, blurring even the distinction between contrast resolution

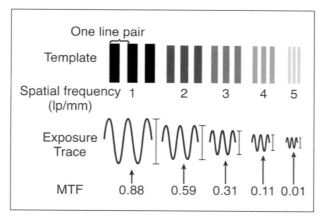

Figure 3-13. The resulting MTF values (bottom) as resolution template line pairs become smaller and closer together. The exposure trace in the middle shows graphically that contrast resolution is lost as the lines become closer. This is also indicated in the simulated image of the lines at the top. (From Q. B. Carroll, *Radiography in the Digital Age,* 3rd ed. Springfield, IL: Charles C Thomas, Publisher, Ltd., 2018. Reprinted by permission.)

and spatial resolution. This is something that can *only* happen at the microscopic level, (not to the clinical image of a bone).

The upshot is that, at the microscopic level of the image, resolution can be lost in two ways: 1) by loss of contrast resolution (the vertical dimension of the exposure trace), or 2) by loss of spatial resolution when small details become so close together that their penumbras overlap.

Conclusion

These microscopic effects discussed in the last section are measured using test devices such as the line-pair template in Figure 3-10 and are useful for comparing the performance x-ray units. However, in daily practice, the radiography is primarily concerned with the "clinical qualities" of the image, the diagnostic value of the gross anatomy and pathology displayed. These are the six qualities listed at the bottom of Figure 3-1: Brightness (density), contrast (gray scale), noise, sharpness, shape distortion and magnification.

A helpful glossary is provided at the end of this book where concise definitions are given for every image quality. This makes a perfect study tool for review.

Chapter Review Questions

1. The higher the window level number, the _____ the image.
2. The *percentage* or *ratio* of difference between two adjacent densities is the definition for image _____.
3. Whereas higher contrast is associated with greater visibility, longer *gray scale* is

associated with more _____ present in the image.

4. The higher the window _____, the longer the gray scale.

5. In the image, gray scale is generally considered as _____ to contrast.

6. For both brightness and contrast, the "ideal" amount for diagnostic image is _____ (maximum, minimum, or optimum).

7. When measuring contrast in a radiographic image, it is essential to take both measurements from _____ within the anatomy.

8. In its broadest sense, *noise* is defined as anything that interferes with the _____ of useful information in the image.

9. Within the remnant x-ray beam, different areas of x-ray intensity possess _____ contrast between them.

10. Within the latent image, the ratio of useful, constructive information to non-useful, destructive input defines the concept of _____ _____ _____ _____ (SNR).

11. If one visually scans across the image, the "abruptness" with which the edge of a particular structure "stops", or how suddenly the structure's density transitions to the background density in the image, is the definition for image _____.

12. In the digital age, what are the two general types of image magnification?

13. Any form of "post-collimation" that changes the field of view displayed on a viewing monitor will affect the _____ magnification of the image.

14. Brightness, contrast and noise impact the *visibility* of an image structure. Sharpness, shape distortion and magnification combine to determine whether the structure can be _____ for what it is.

15. With the single exception of shape distortion, every aspect of the latent radiographic image will be modified by _____ processing before the image is displayed.

16. For a physicist studying the microscopic resolution of a single dot, there are only two aspects to the image, _____ resolution which best corresponds to the *visibility* of the dot, and _____ resolution which best corresponds to the overall *recognizability* of the dot.

17. On an exposure trace diagram, the _____ depth of a "pit" in the trace corresponds to the contrast of the detail.

18. On an exposure trace diagram, the horizontal extent of the detail's edge slopes corresponds inversely to the _____ of the detail.

19. For a line-pair test image, loss of spatial resolution can occur when smaller and smaller lines become so close that their _____ begin to overlap.

20. List the six essential qualities of any *gross* anatomical image (such as an image of a bone) that radiographers are concerned with in daily practice:

Chapter 4

RADIOGRAPHIC TECHNIQUE FOR DIGITAL IMAGING

Objectives

Upon completion of this chapter, you should be able to:

1. State the primary role of radiographic technique in the digital age.
2. Describe how x-ray beam penetration distinguishes between exposure to the patient and exposure to the image receptor.
3. Compare the sensitivity to subject contrast between digital and film technology
4. Explain the relationship between subject contrast and exposure latitude.
5. Describe the exposure latitude of digital equipment.
6. For reduced use of grids, compare the implications for scatter radiation to the occurrence of mottle, and the implications for setting radiographic technique.
7. State three advantages for the use of new *virtual grid* software.
8. Describe how long gray scale in the latent image maximizes the quantity of real data.
9. Quantify how the 15 percent rule reduces patient exposure while maintaining exposure to the image receptor.
10. Analyze the impact of applying the 15 percent rule on digital image "fog" and digital image mottle.

If, as described in the previous chapter, the qualities of the final displayed radiographic image are now primarily controlled by digital processing, then what role is left for the radiographic technique set at the console during exposure? The answer is both significant and singular: *To provide adequate signal at the image receptor for the computer to be able to manipulate the data.* That is, to ensure that adequate organized information reaches the image detector along with minimal noise, or to achieve the maximum SNR (signal-to-noise ratio).

Of course, another important goal in the practice of radiography is to minimize patient dose, and this also relates to the set radiographic technique; However, this will be discussed later. Here, we want to focus on production of a quality image.

Without proper radiographic technique, the digital image would not exist. The computer must receive sufficient input so that it can then "crunch the numbers" to produce a diagnostic final image. Positioning, focal spot size and x-ray beam geometry still have a major impact upon the *recognizability* aspects of the image (sharpness, distortion and magnification), but what would any of these matter if the image is not sufficiently *visible,* i.e., it doesn't exist? Enough data must be fed into the computer that it can, in effect, parse it and select the most pertinent data, move it up or down along a dynamic range of brightness levels, and display the result with sufficient contrast and minimal noise.

Since an adequate amount of the x-ray beam must make it through the various structures of the patient's body (and through other necessary objects such as the tabletop and the grid) to finally reach the image receptor, our primary concern for setting radiographic technique is sufficient x-ray *penetration.*

Understanding X-Ray Beam Penetration

X-ray beam penetration may be generally defined as the *percentage* or *ratio* of x-rays that make it through the patient, tabletop and grid to strike the image receptor. Suppose that, for simplicity's sake, we begin with a primary x-ray beam containing 100 x-rays, and 10 percent of them make it through the patient to the image receptor. The total exposure at the image receptor is 10 x-rays. If we can double the penetration to 20 percent, Figure 4-1, then we need only start out with 50 x-rays emitted from the tube to achieve the same exposure at the image receptor (10% of 100 = 20% of 50). Either approach will result in an exposure to the IR of 10 x-rays.

At the console, the primary control for penetrating "power" of the x-ray beam is the kVp (kilovoltage-peak). Although increasing the kVp also adds a bit to the amount of x-rays produced, this quantity is affected much more by the mAs (milliampere-seconds) set at the console, in fact, it is directly proportional to the mAs. From the above example illustrated in Figure 4-1, we conclude that the *total exposure at the image receptor is not based on the mAs alone, but on the combination of kVp and mAs used.*

All the image receptor "cares about" is its exposure level from the *remnant* x-ray beam reaching it behind the patient, not so much the original amount of x-rays produced in the primary beam by the x-ray tube. This total exposure at the IR depends *both* upon the original quantity *and* penetration through the patient. It is thus impacted by both mAs and kVp.

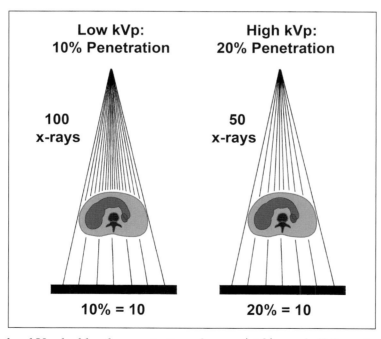

Figure 4-1. If using higher kVp doubles the penetration of x-rays (*right*), one-half the mAs can be used to deliver the same exposure to the image receptor.

Sensitivity of Digital Units to Subject Contrast

Subject contrast is the ratio of difference in radiation intensity between adjacent areas of the remnant x-ray beam that represent different tissues within the body. It is produced by differences in body part thickness, the average atomic number of different tissues, and the physical density (mass/volume) of different tissues.

For any image to be produced, a threshold level of subject contrast, that is, a certain minimum of difference between these tissue areas, must exist within the data provided to the computer; Otherwise, effectively, the computer might not be able to distinguish between one density and another, just as with the human eye. What is this minimum level of subject contrast? And, how has it changed with the advent of digital imaging?

The minimum subject contrast that is required varies with different imaging modalities. For example, note that white and gray matter in the brain can be distinguished from each other on a CT image, but not on a radiograph (Figure 4-2). The CT scanner is *more sensitive* to the original subject contrast present in the incoming image, so it is able to display smaller differences between tissues in the final CT image.

With conventional film radiography, a subject contrast difference of at least 10 percent between adjacent tissues was necessary to distinguish between them in the final film image. Because of the contrast-enhancing capabilities of digital imaging software and the sensitivity of digital detectors, it is now possible to demonstrate in the final image tissues with as low as 1 percent subject contrast (see Figure 4-3). This is *10 times* the contrast resolution of film technology.

Subject Contrast and Exposure Latitude

Exposure latitude is the range of different radiographic techniques that can be used for a particular exposure and still produce a diagnostic image of acceptable quality. Exposure latitude may be thought of as the *margin for error* in setting a radiographic technique. In this context, "technique" includes not only the exposure factors set at the control console, but also the use of different grids, filters, focal spots and distances that can affect the image.

Generally, *a latent image possessing higher subject*

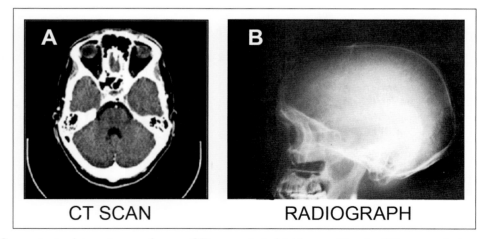

CT SCAN   RADIOGRAPH

Figure 4-2. Computerized imaging, such as a CT scan, *A*, is highly sensitive to differences between tissues (see Figure 4-3), demonstrating the eyeballs within periorbital fat, and both the gray and white matter of the brain tissue. A conventional radiograph of the skull, *B*, demonstrates only bone, air, and fluid-density tissues. (From Q. B. Carroll, *Radiography in the Digital Age*, 3rd ed. Springfield, IL: Charles C Thomas, Publisher, Ltd., 2018. Reprinted by permission.)

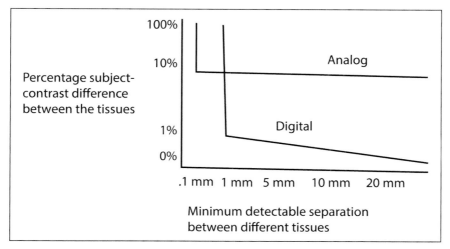

Figure 4-3. Digital systems can demonstrate tissues with less than 1 percent inherent subject contrast. Analog imaging, such as conventional film imaging, was unable to demonstrate any tissues with less than 10 percent subject contrast between them. (From Q. B. Carroll, *Radiography in the Digital Age,* 3rd ed. Springfield, IL: Charles C Thomas, Publisher, Ltd., 2018. Reprinted by permission.)

contrast will present less exposure latitude, or less margin for error. This is graphically illustrated in Figure 4-4. As described in the previous chapter, high contrast is generally associated with *short gray scale, A* in Figure 4-4. The length of the solid arrow represents those densities actually being displayed, (To simplify the illustration, this number has been reduced from a realistic range of densities.) Note that, effectively, there is *less room* to move this range up or down before running into the extremes of blank white on one end or pitch black on the other. Yet on the right, *B,* where the gray scale is lengthened, the exposure level can be moved up or down by the same amount without "running out of densities." This implies that there is more leeway to increase or decrease technique without gross underexposure or overexposure occurring.

When only a 1 percent difference in subject contrast is required for digital equipment to resolve a good final image (previous section), the result is a much extended exposure latitude. There is generally more margin for error, and along with it, more latitude for experimentation and innovation in setting radiographic technique.

One result is the flexibility to employ much higher kVp levels than were required with conventional film radiography. Another is that the

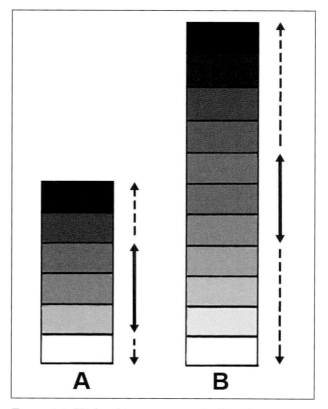

Figure 4-4. High subject contrast, *A*, allows less *exposure latitude.* There is less "room" in image *A* to move the range of displayed densities (*solid arrow*) up or down before running into blank white or pitch black extremes. Long gray scale, *B*, allows more margin for error.

use of particular grids and filters during the exposure is *less critical* (not more critical, as some publications have suggested)—it is widely accepted that digital radiography has much wider exposure latitude than film radiography had. A generally wider margin for error can only mean that *all* technical aspects of the original exposure become less critical. This allows us to be more inventive with all aspects of radiographic technique.

An important footnote must be made at this point, however: The increased exposure latitude of digital radiography extends primarily in an *upward* direction, toward overexposure, not so much downward. This is because with digital equipment, when the total exposure is reduced very much below one-half of the ideal amount for the anatomy, mottle may begin to appear in the final image. At exposures less than one-third the ideal, unacceptable levels of mottle are certain to be present. So, when we conclude that we can be more innovative with technique, we mostly refer to changes that will result in *more* exposure reaching the IR. For these kinds of changes, the only other restricting factor is their effect on patient dose, which must be taken into consideration.

Use of Grids

The proper use of grids has become a somewhat controversial subject for digital equipment, complicated by the fact that grids can have a substantive impact on the occurrence of mottle (a form of noise), scatter radiation (another form of noise), and patient dose. All three of these effects must be weighed against one another.

First, let's reiterate from the previous section that the greatly-widened exposure latitude of digital radiography allows *more flexibility* in the use of grids. The immediate possibilities that arise are:

1. Using *non-grid* techniques for some procedures that used to require grids
2. Using *lower grid ratios* for procedures that still require a grid

Note that both of these practices would allow less mAs to be used, which reduces radiation exposure to the patient, and are thus well worth pursuing. Generally, a non-grid technique requires about one-third (or less) of the mAs needed for a grid exposure. Patient exposure is directly proportional to the mAs used, so "going non-grid" means reducing patient exposure to one-third or less, a dramatic savings in dose.

Where a 6:1 ratio grid can be used instead of a 10:1 or 12:1 ratio, the mAs can be cut approximately in half. These savings in patient dose are so substantial, especially compared to other recommended "dose-saving" measures, that every radiography department should take them into serious consideration wherever they can be applied. Some hospitals have already made these changes and, justifiably, advertised their "low-dose" policies for PR purposes.

What are the implications of such changes for image quality? Grids were invented to reduce the amount of scatter radiation reaching the image receptor, which was substantial. Scatter is a form of noise and reduces the signal-to-noise ratio (SNR) at the detecting device. By changing the dexel values (pixel values) fed into the computer, it corrupts the image data somewhat. However, default digital processing routinely amplifies the contrast of the image, restoring nearly all "damage" done by moderate amounts of scatter radiation. (Remember that only a 1 percent subject contrast is now needed, so if scatter radiation reduced the subject contrast from 10 percent to 5 percent, the computer can still differentiate between these tissue areas and successfully process the image. The computer can even compensate for expected "fogged" portions of images such as tend to occur on a lateral lumbar spine, which will be discussed in Chapter 6.) *Only the most extreme amounts of scatter show up in the quality of the final displayed digital image.*

Quantum mottle is another form of image noise, and shows up as a "speckled" appearance within image densities. Quantum mottle results from an uneven distribution of radiation within the x-ray beam, a natural statistical variation in the amount of radiation intensity from one area to another. It becomes visible in the image when a very low amount of exposure to the image receptor leads to insufficient signal. Just as individual raindrops can be seen on a sidewalk during a light shower, but disappear during a downpour, quantum mottle also disappears from view when plenty of x-rays "flood" the image receptor with good signal.

Grids require higher techniques because their lead strips absorb not only scatter radiation from the patient, but also part of the primary radiation originally emitted by the x-ray tube. This loss of exposure is called grid *cut-off.* If the technique were not increased to compensate, less radiation would reach the image receptor. If exposure to the IR is lowered enough, visible quantum mottle may appear in the image. Digital processing has noise-reduction algorithms that can correct for moderate amounts, but not extreme amounts, of mottle in the input image.

In fact, several manufacturers now feature "virtual grid" algorithms that literally replace the need for grids except in the most extreme situations, such as abdominal radiography on obese patients. The virtual grid program can be applied at the touch of a button, or may be included in the default programming so it is automatically applied for selected anatomical procedures.

Figures 4-5 and 4-6 demonstrate the very impressive effectiveness of *Fuji's Virtual Grid*™ software in cleaning up "scatter effect." Figure 4-5, an AP projection of the pelvis, shows the difference when virtual grid software is applied to a non-grid image. More to the point, Figure 4-6 compares a conventional grid image to a virtual grid image, using a PA chest projection. This comparison is representative of most manufacturers' versions of virtual grid software: As we might expect, using software to correct for scatter is not 100 percent as effective as using a conventional grid. However, both Figures 4-5 and 4-6 confirm that *virtual grid software is about 85 percent as effective as conventional grids, and given the advantages to both patient dose and to positioning, the case for changing from conventional grids to virtual grid software is compelling.*

Removing a grid, or reducing the grid ratio

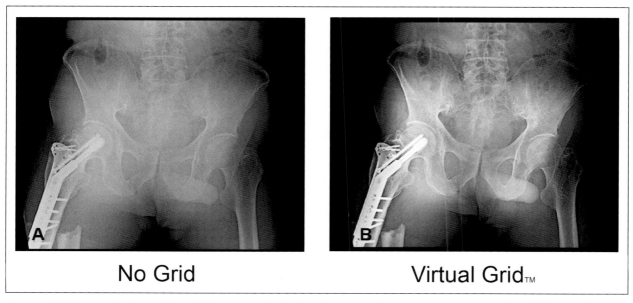

No Grid   **Virtual Grid**™

Figure 4-5. Demonstration of the effectiveness of *virtual grid* software in cleaning up scatter effect. Left, a pelvis projection taken without using VG software and with no conventional grid. Right, Virtual Grid™ software applied with no conventional grid used. (Courtesy of FUJIFILM Medical Systems, U.S.A, Inc. All rights reserved. Reproduced with permission.)

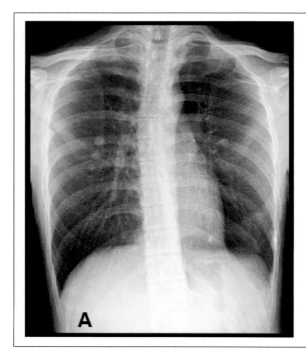 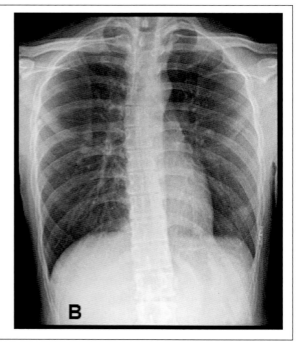

Figure 4-6. Comparison of the effectiveness of a conventional 12:1 grid, **A**, with *Virtual Grid™* software applied to the same chest projection and no conventional grid used during exposure, **B**. (Courtesy of FUJIFILM Medical Systems, U.S.A, Inc. All rights reserved. Reproduced with permission.)

used, allows more radiation to reach the image receptor and thus contributes to *preventing* the appearance of mottle. On the other hand, removing a grid or reducing the grid ratio allows *more* scatter radiation to reach the IR. One type of noise is reduced, the other increased, for the latent image input into the computer. In terms of the overall SNR (signal-to-noise ratio), these two effects largely cancel each other out.

But, not *completely:* In practice, scatter radiation is somewhat more likely to present a problem. For this reason, we cannot entirely eliminate the use of grids "across the board." We have stated that removing or reducing the grid, by allowing a reduction in technique, can substantially reduce patient exposure, which must still be weighed against this slight increase in noise in the form of scatter radiation. Let's summarize the considerations for removing or reducing grids with digital equipment:

1. The effects of scatter radiation are increased slightly

2. The likelihood of mottle appearing is reduced

3. Grid cut-off is eliminated, allowing more flexibility in positioning

4. Radiation exposure to the patient is reduced

As you can see, in the overall practice of radiography, the benefits of removing or reducing grids outweigh the disadvantages considerably. The ability of digital processing to tamper down the effects of scatter radiation has made this possible. The following clinical applications are suggested:

1. Consider non-grid techniques for:
   Knees
   Shoulders
   C-Spines (except cervico-thoracic)
   Sinuses/Facial bones
   All pediatrics

2. Use a low, 6:1 grid ratio for all gridded mobile procedures

3. Consider reducing grid ratio to 6:1 or 8:1 in all fixed radiographic units

The first recommendation relates to a conventional rule that "grids should be used on any anatomy exceeding 10 cm in thickness." *With the increased exposure latitude of digital technology, this threshold should be raised to 13 cm.* For example, it is not necessary to use a grid on most adult knees. Actual measurements will reveal that the average adult knee falls between 11 cm to 13 cm in thickness. The upper *tibia* averages 9-10 cm, so the old "10 cm rule" is exaggerated, implying that anything larger than the leg requires a grid.

Sufficient Input Gray Scale

Described in Chapter 6, the computer is able to effectively lengthen or shorten the gray scale

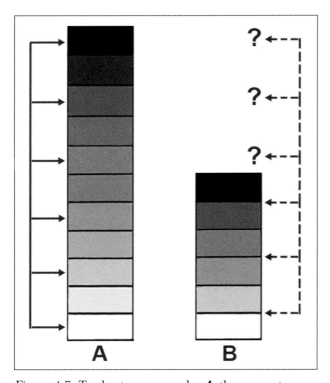

Figure 4-7. To shorten gray scale, *A*, the computer can sample known data (every other density). But, to lengthen gray scale that is too short, *B*, the computer must *interpolate* between densities, effectively fabricating information.

that will eventually be displayed in an image, through *gradation processing.* The range of different pixel values available to the displayed image is increased or decreased mathematically. But, let's consider the implications of each process:

Suppose that from the latent image recorded at the detector plate, a long scale of gray shades (dexel values or pixel values) is fed into the computer, but it is desired to shorten the gray scale for the final displayed image. Shown in Figure 4-7, one way this could be done is to simply use every other density or gray scale value from the original input. In this case, all of the densities used would be *real values as measured at the image detector.*

Now, imagine that the latent image at the detector plate consists of a very short gray scale so that, in effect, only a handful of dexel values or pixel values is made available to the computer, but it is desired to the *lengthen* the gray scale for the final displayed image. The computer can do this only by a process called *interpolation,* "filling in the blanks" by finding numbers that are mid-way between those provided from the detector. The important point is that these are *not real, measured dexel or pixel values,* they are values *fabricated* by computer algorithms and as such, constitute *artificial information.* The process works in terms of "filling out" more gray scale for the image, but it must be remembered that the image now contains pixel values that were not actually measured and may not accurately represent actual tissues in the body.

When sufficient gray scale is originally inputted to computer, that scale can be reduced without any misrepresentation of actual densities. On the other hand, when too short a gray scale is inputted into the computer, the scale can only be lengthened by effectively *fabricating* new and artificial density values. Long gray scale in the latent image is primarily achieved through *high kVp techniques.*

Minimizing Patient Exposure with the 15 Percent Rule

As described at the beginning of this chapter, what matters at the image receptor is sufficient exposure, simply the total count of x-rays reaching it through the patient, and this total exposure depends on *both* the mAs and kVp set at the console. Through the decades, the *15 percent rule* has been used commonly by radiographers; It adjusts the mAs for changes in kVP such that the end result is approximately the same total exposure to the IR.

Since mAs is directly proportional to patient exposure, cutting the mAs in half also results in cutting patient dose in half, a desirable outcome for the patient. To compensate, a 15 percent increase in kVp restores the original exposure to the IR. There are *two* reasons this works: First, although fewer x-rays are being emitted in the primary beam, a higher *percentage* of them penetrate through the patient to the IR. This recovers about two-thirds of the original total exposure to the IR.

Second, a 15 percent increase in kVp actually results in, on average, about a 35 percent increase in the real number of x-rays being produced in the x-ray tube anode. Specifically, these are *bremsstrahlung* x-rays, sometimes called "braking radiation" because they are emitted

each time a projectile electron passes near an atomic nucleus and is "pulled" by the positive charge of the nucleus, slowing the electron down such that it loses kinetic energy. This is analogous to "putting on the brakes in your car." The speed energy lost by the electron is emitted as an x-ray.

When an electron possessing higher kV strikes the anode, it is literally traveling faster to begin with. Whereas a slower electron may have required three encounters with atomic nuclei to use up its speed energy, this high-kV electron may have to be "braked" *five times,* by five atomic nuclei, before its speed energy is used up. Thus, the high-kV electron produced five x-rays, while the lower-kV electron only produced three x-rays.

Now, let's combine the two effects from applying the 15 percent rule: Cutting the mAs in half reduces the original exposure to 50 percent, while the increase in kVp adds back 35 percent of the 50. Thirty-five percent of 50 is 17, adding 17 to 50 we get 67 percent as an end result. This is the amount of radiation the surface of the patient receives. *Applying the 15 percent rule to in-*

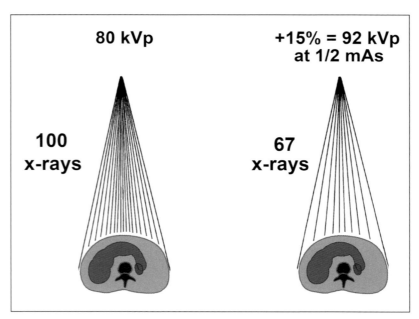

Figure 4-8. Cutting the mAs in half reduces x-rays in the primary beam to 50 percent, but increasing the kVp by 15 % adds about 1/3 more bremsstrahlung x-rays back in. The net result for patient exposure is 50% + 17% (1/3 of 50) = 67% total.

crease kVp with mAs compensation results in a reduction in patient exposure to about two-thirds the original exposure, Figure 4-8. Patient exposure has been cut by a third.

What happens at the image receptor is completely different, because now the aspect of *beam penetration* enters into the equation. (Penetration through the patient was irrelevant at the patient's surface, but at the IR, *behind the patient,* it is now quite relevant.) In Figure 4-9, we show (at the bottom) an increase in penetration from 6.7 percent to 10 percent when the kVp was raised from 80 to 92. When the increase in the percentage of x-rays now penetrating through the patient is added to the increase in *bremsstrahlung* production, the end result is that *exposure to the IR is completely restored (back to 100 percent).* In Figure 4-9, *both* image receptors receive 67 x-rays.

To summarize, as illustrated in Figure 4-9, *each time the 15 percent rule is applied to increase kVp and compensate mAs, patient exposure will be reduced by about one-third while exposure to the IR is maintained.*

Benefits of High kVp Radiography

With digital imaging, higher kVp techniques can be used than with conventional film radiography because the minimum subject contrast needed between tissues is now only 1 percent. The *benefits* of using a high-kVp approach to radiographic technique may be summarized as follows:

1. It helps ensure sufficient x-ray penetration through the patient to the IR
2. It provides long gray scale input to computer for manipulation without interpolation
3. It reduces patient exposure

The greatly lengthened exposure latitude for digital radiography has removed nearly all restraints from high-kVp practice. For conventional film radiography, the main restriction arose from the risk of producing "fog" densities on the radiograph from scatter radiation. Figures 4-10 and 4-11 demonstrate that this concern has

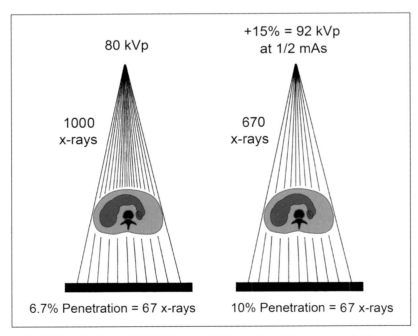

Figure 4-9. Each time the 15% rule is applied to increase kVp and cut the mAs in half, patient dose is reduced by about 1/3 (to about 67%), yet exposure to the image receptor is maintained because of increased penetration through the patient to the IR (here, from 6.7% to 10% as a simplified example).

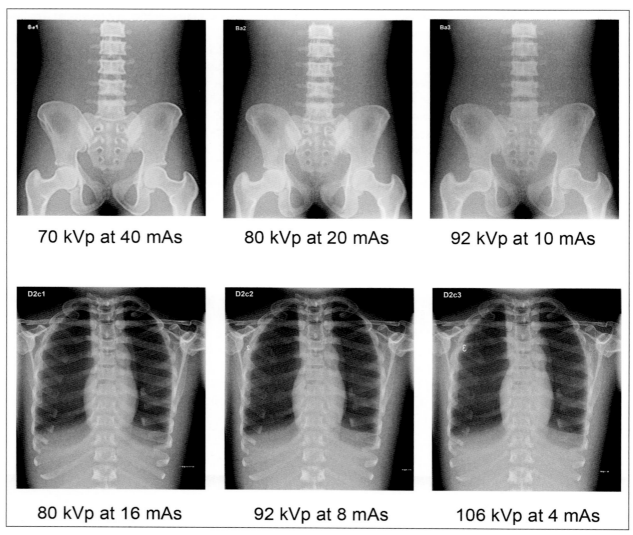

70 kVp at 40 mAs   80 kVp at 20 mAs   92 kVp at 10 mAs

80 kVp at 16 mAs   92 kVp at 8 mAs   106 kVp at 4 mAs

Figure 4-10. Series of abdomen and chest phantom digital radiographs, taken with 15% step increases in kVp and the mAs cut to 1/2 for each step, demonstrate that for any one step increase of 15% kVp, the change in contrast between any two adjacent radiographs is so slight that the difference can barely be made out visually. *No fog densities are present.* The original study proved this to be true over *5 step increases* in kVp with consistent results for *nine different manufacturers.* (Courtesy, Philip Heintz, PhD, Quinn Carroll, MEd, RT, and Dennis Bowman, AS, RT. Reprinted by permission.)

been essentially removed from the practice of digital radiography. The overall reduction in image contrast is *visually negligible* for each 15 percent step applied. It is so minor that, for most digital equipment, it takes several steps to see any change. Furthermore, even if a difference is noted, contrast can be instantly adjusted back upward as desired at the display monitor, something radiologists routinely adjust to personal preference anyway.

Not only can digital processing deal with a much lower subject contrast in the inputted image, but there are also computer algorithms that can identify and correct for expected fog patterns such as those encountered on the lateral lumbar spine projection (Figure 4-12).

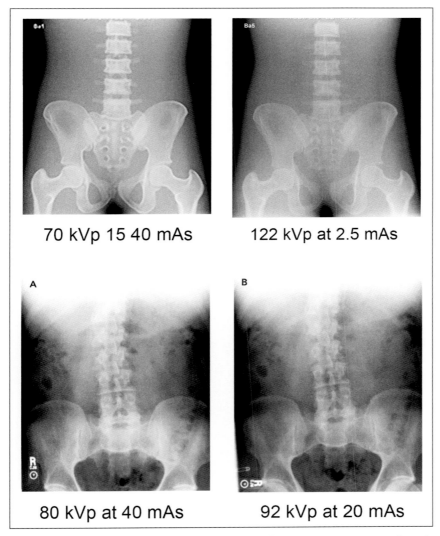

70 kVp 15 40 mAs   122 kVp at 2.5 mAs

80 kVp at 40 mAs   92 kVp at 20 mAs

Figure 4-11. A *52 kVp* increase between two digital radiographs (*top*) shows about the same lengthening of gray scale as a *single* 15% increase in kVp on conventional film radiographs (*bottom*). *There is no fog pattern in the high-kVp digital image. Lengthened gray scale is from dramatically increased penetration.* (Courtesy, Philip Heintz, PhD, Quinn B. Carroll, MEd, RT, and Dennis Bowman, AS, RT. Reprinted by permission.)

The only remaining concern for applying the 15 percent rule is whether the attendant reductions in mAs would be additive, and the rule-of-thumb itself being slightly inaccurate, such that a repeated sequence of 15 percent steps might result in quantum mottle becoming apparent. Figure 4-13 displays some of the resulting images from an extensive study done on this question. The experiment was repeated for *nine* different manufacturers of digital x-ray equipment. For each sequential step applied, the kVp was increase by 15 percent while the mAs was cut in half. The conclusion, supported by radiologists, was as follows:

1. The *third* step-increase generally resulted in a visible increase in mottle
2. *All but one* brand of digital equipment showed no appreciable mottle for two steps applied
3. Mottle was never significant for a single-step application of the 15 percent rule

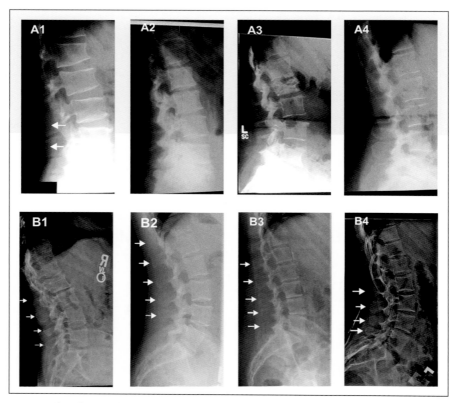

Figure 4-12. Conventional radiographs of the lateral lumbar spine (*top row*) typically showed a distinct fog pattern (*arrows*) obscuring the ends of the spinous processes. Digital processing nearly always demonstrate the full spinous processes (*bottom row*), having removed most of the fog pattern. (From Q. B. Carroll, *Radiography in the Digital Age,* 3rd ed. Springfield, IL: Charles C Thomas, Publisher, Ltd., 2018. Reprinted by permission.)

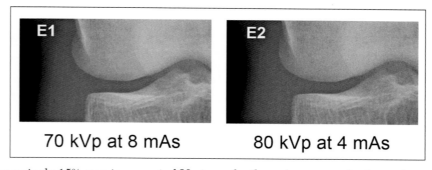

Figure 4-13. When a single 15% step increase in kVp is made, the mAs can nearly always be cut to one-half without the appearance of mottle (*right*). This was proven to be true for nine manufacturers. (From Q. B. Carroll, *Radiography in the Digital Age,* 3rd ed. Springfield, IL: Charles C Thomas, Publisher, Ltd., 2018. Reprinted by permission.)

Table 4-1
RECOMMENDED OPTIMUM KVP FOR DIGITAL IMAGING

Recommended Optimum kVp for Digital Imaging			
Procedure	**Optimum kVp**	**Procedure**	**Optimum kVp**
Hand/Wrist/Digits	64	Iodine Procedures (IVP, cystog)	80
Elbow/Forearm/Foot	72	Abdomen/Pelvis/Lumbar Spine	90
Ankle/Leg	76	Non-Grid Chest (fixed unit)	86
Knee/Humerus Tabletop	80	Skull	90
Knee in Bucky	84	Air Contrast Barium Studies	100
Femur/Shoulder/Sinus/Ribs	86	Esophagram	92
Mandible, Tangential Skull	76	Solid-Column Barium Studies	120
Cervical and Thoracic Spines	86	Grid Chest (fixed unit)	120
Pediatric Extremities	60-70	Pediatric Chest	70-80

This means that any x-ray department should be able to at least make a single 15 percent increase in kVp *across the board,* that is, for all technique charts, and cut all mAs values to one-half. By doing so, *radiation exposure to patients would generally be cut by one-third while image quality was maintained.*

Because of the exposure latitude of digital radiography, there is simply no compelling reason to keep kVp at the levels used for film technology, when we were constrained by the general fogging effects of scatter to define a *minimum kVp* recommended for each body part. Digital technology has eliminated the general concern for scatter, allowing us to focus on saving patient dose. We now define the *optimum kVp* as that level of kVp that strikes an appropriate balance between reducing patient dose and preventing excessive scatter radiation at the IR. Based upon this definition, Table 4-1 presents a listing of optimum kVp's for various landmark body parts. *These optimum kVp's are strongly recommended for clinical use in digital radiography.*

Chapter Review Questions

1. In the digital age, the primary role for the set radiographic technique is to provide sufficient _____ at the image receptor for the computer to be able to manipulate the data.
2. The percentage or ratio of primary beam x-rays that make it through the patient is the definition for _____.
3. The total exposure to the image receptor is not based on the set mAs alone, but upon the combination of _____ and _____ set at the console.
4. For any image to be produced, a _____ level of subject contrast must exist within the remnant x-ray beam

5. Digital imaging equipment, such as a CT scanner, can demonstrate tissues with as low as _____ percent subject contrast.

6. This ability is _____ times the ability of film technology to demonstrate low-subject contrast tissues.

7. For a particular procedure, the range of techniques that can produce acceptable quality image, or the margin for error in setting technique, is called _____ _____.

8. For digital equipment, very high exposure latitude means that all technical aspects of the original exposure become _____ critical, allowing the flexibility to reduce grid use and employ high kVp.

9. Generally, a latent image possessing higher subject contrast will present _____ exposure latitude.

10. However, this increased exposure latitude extends primarily in a(n) _____ direction.

11. An exposure less than _____ the ideal is certain to present unacceptable levels of mottle.

12. Which image problem is more difficult for digital technology to compensate for, scatter radiation or mottle?

13. State three advantages of reducing grid use or grid ratio:

14. *Virtual grid* software is about ____ percent as effective as conventional grids in reducing the effects of scatter radiation.

15. If the latent image has high subject contrast, and longer gray scale is desired, the computer must _____ new density values in the image that are not based on real, measured data.

16. The 15 percent rule works in reducing patient dose because, even though a 15 percent increase in kVp causes 35 percent more x-rays to be initially produced, cutting the mAs to one-half cuts patient dose to _____, (a greater change).

17. The net result of applying the 15 percent rule is that patient dose is cut to about _____ _____, while exposure to the image receptor is _____.

18. State three benefits of high-kVp radiography:

19. With digital technology, for a single 15 percent step increase in kVp, the overall reduction in image contrast is visually _____.

120 It has been demonstrated that with nine different brands of digital technology, for a single step increase of 15 percent in kVP, image mottle was _____ significant.

Chapter 5

PREPROCESSING AND HISTOGRAM ANALYSIS

■ ■

Objectives

Upon completion of this chapter, you should be able to:

1. List the generic steps for *preprocessing* and *postprocessing* digital radiographs.
2. Describe corrections for field uniformity, dead dexels, random mottle and periodic mottle.
3. Explain how the acquired image histogram is constructed, and interpret the graph.
4. Describe threshold *algorithms* and how the computer "scans" the histogram for landmarks.
5. Describe how the *lobes* of histograms are used as landmarks for analysis.
6. Explain the significance of the SMAX point, the SMIN point, and the values of interest (VOI) in histogram analysis.
7. Describe the types of circumstances that lead to histogram analysis errors.

For about one hundred years, from the discovery of x-rays until the digital revolution near the turn of the 21st century, radiographs were processed chemically. Early photographs and radiographs were recorded on glass plates coated with a silver compound; Exposure to light or x-rays essentially tarnished the silver. When the plate was then immersed in a liquid developer solution, chemicals would attack each exposed silver molecule, continuing the "tar-nishing" process until the molecule was completely black. Unexposed crystals had no breaks or "cracks" in their molecular structure, and resisted development by not letting the chemicals penetrate. This left unexposed portions of the image clear or "white." To produce various shades of gray in the final image, where some proportion of the x-rays had penetrated through to the plate, different percentages of developed black crystals would be mixed in with unexposed white crystals to varying degrees.

Plastic film bases quickly replaced the early glass plates which were both fragile and heavy. Coated with silver compounds on both sides, plastic film was more responsive to exposure, durable, light and flexible so it could be hung on viewbox light panels for diagnosis. In the digital age, computer-processed images can still be printed out onto clear plastic film to produce permanent copies.

In effect, radiographic film measured exposure or "counted" x-rays by how dark different areas of the image turned out upon chemical development, that is, how much chemical change had occurred. In all modern digital x-ray systems, the image receptor is an *electronic* detector rather than a chemical detector—Exposure is measured (x-rays are "counted") by how much electrical charge is built up. Both processes, old and new, are based on the ability of x-rays to *ionize* atoms, knocking electrons out of their atomic orbits. With film, these ionizations led to chemical changes that looked darker;

With digital equipment, the freed electrons are stored up on a capacitor, somewhat like a miniscule battery, and the amount of charge is then measured. Each measurement is recorded as a "pixel value," and all these numbers together make up the data set that will be processed by a computer

Conventional chemical processing of films consisted of only four steps—development, fixing, washing and drying. Modern digital image processing consists of some nine generic steps, after reducing and simplifying as much as possible.

The Generic Steps of Digital Image Processing

By way of introduction to digital processing, an important general distinction must be made: For traditional chemical processing of films, the four steps were also the *sequence* and absolutely had to be followed in order. One could not chemically fix a film that had not first been developed, or dry one that had not yet been washed, to end with the desired result. For digital processing, the nine listed steps can be applied in different order, and the various manufacturers of digital equipment do so in a number of ways. Furthermore, some steps are repeated at different stages of processing two or even three times.

An excellent example is the generic step called *noise reduction.* Early in the preprocessing stage of a DR image, it is desirable to correct for *dead* pixels in the image, where detector elements in the image receptor have electronically failed and "blank" spots are left in the image. Such blank spots are rightly considered as a form of noise, and noise reduction software is used to fill each of them with an appropriate density. Later as part of the default process, *detail processing* that includes edge enhancement often leaves the image with an unnatural, "alien" appearance. The image is restored to a more normal look by *smoothing* algorithms that are one and the same as noise reduction. Finally, after the image is displayed, the operator may choose to apply a "softer look" to the image at the push of a button. This operation also falls

under the general category of *smoothing* and uses the same algorithms as noise reduction, the *third* time the same software package will have been applied.

We can broadly divide the nine steps of digital image processing into two categories: preprocessing and postprocessing. *Preprocessing* is defined as all computer operations designed to *compensate* for *flaws in image acquisition,* corrections that are necessary to give the digital image the appearance of a conventional radiograph. Without these corrections, the image is both noisy and so extremely "washed-out" (Figure 5-1) that it doesn't even look like a typical black-and-white "picture" and is not of any diagnostic value.

Postprocessing is defined as all those steps that might be considered as *refinements* to the image after it has reached a stage where it has the appearance of a typical black-and-white picture. These refinements are customized according to each specific radiographic *procedure,* to produce image traits that best demonstrate that particular anatomy and the common types of pathology that occur within that anatomy. *Postprocessing is targeted at the specific anatomical procedure, whereas preprocessing is targeted at basic image acquisition.* The nine steps of digital image processing are listed under these categories in Table 5-1.

Now, some confusion might arise from these terms, since *pre*processing implies that the step takes place *before* processing, and *post*processing indicates that the step occurs *after* processing. One might legitimately ask, if preprocessing and postprocessing cover all nine steps, whatever happened to just *processing?* The digital processing step that most closely approximates this concept is step number 4 in Table 5-1, *rescaling.* It is during rescaling that the digital image is given the appearance of a typical black-and-white picture—this is when black, dark gray, light gray, and white pixel values are assigned to each pixel such that the image takes on the contrast necessary for diagnosis. Therefore, in Table 5-1, *processing* has been added in parentheses next to the rescaling step.

The problem is that rescaling is made necessary because of a *flaw* in digital image acquisi-

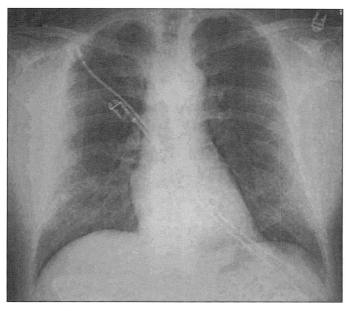

Figure 5-1. "Raw" latent image of a PA chest as recorded by a DR detector and prior to any digital processing. (From Q. B. Carroll, *Radiography in the Digital Age*, 3rd ed. Springfield, IL: Charles C Thomas, Publisher, Ltd., 2018. Reprinted by permission.)

tion, namely, that the "raw" digital image coming from the image receptor has such poor contrast that it cannot even be diagnosed, as shown in Figure 5-1. This falls under *preprocessing* as we have defined it, and we are compelled to keep our definition of preprocessing because it is the only clear way to distinguish it from postprocessing. Both terms have been in use for some time now by physicists and manufacturers, and this is the simplest approach for presenting these terms to new students of radiography.

Table 5-1
GENERIC STEPS IN DIGITAL PROCESSING

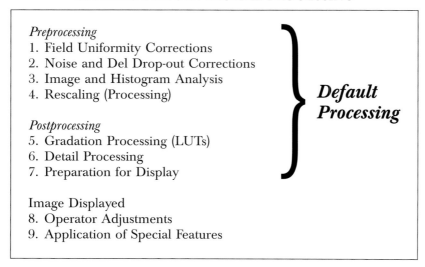

Preprocessing
1. Field Uniformity Corrections
2. Noise and Del Drop-out Corrections
3. Image and Histogram Analysis
4. Rescaling (Processing)

Postprocessing
5. Gradation Processing (LUTs)
6. Detail Processing
7. Preparation for Display

Image Displayed
8. Operator Adjustments
9. Application of Special Features

} ***Default Processing***

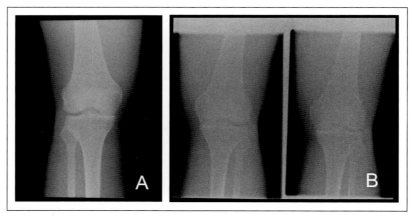

Figure 5-2. *Segmentation error* resulted in image **B** being processed too dark when the strip between the two images was interpreted as anatomy. The light strip across the top was also caused as a processing artifact. Image **A** is provided for comparison.

Preprocessing

In computed radiography (CR), more than one image can be recorded on a single PSP plate (photostimulable phosphor plate). The first order of business for the processing computer is to sort out how many separate images are on the plate, and identify where each one begins and ends so that they are not all processed together as a single image. If they were, the blank spaces between them would be interpreted by the computer as if they were bones or metallic objects within the body part. The computer would interpret a large percentage of the image as being very light and attempt to correct by darkening up the overall image it "sees." This would result in the exposed areas coming out too dark, (Figure 5-2). This type of computer error is commonly known as *segmentation failure,* the inability to segment or separate individual exposure areas, and applies only to CR. Since DR units allow only one exposure on each detector "plate," which is sent to the computer before another exposure can be taken on the same plate, the segmentation step is unnecessary.

Correcting for Dexel Dropout

In the second step for digital preprocessing, the computer looks for "dead pixels," where individual detector elements (dexels) in the image receptor might not report any data due to electronic failure. With many hundreds of small, delicate detector elements in a typical array for a DR image receptor, there are bound to always be a handful that fail. To avoid blank spots showing up in the image the computer uses *noise reduction software.* A great example is a spatial processing tool called a *kernel.* A kernel may be defined as a sub-matrix that is passed over the larger image executing some mathematical function on the pixels. Figure 5-3 illustrates a simplified nine-cell kernel for correcting individual dexel drop-out. This kernel reads and averages the pixel values for the eight pixels surrounding the "dead" pixel, and then artificially inserts that number into the dead spot. This process is known as *interpolation.*

Correcting for Mottle

All digital images have some degree of mottle in them. Mottle is a form of *noise* that manifests as a grainy appearance to the image, consisting of very small freckle-like blotches of dark and light throughout the image (Figure 5-4). Radiographers are particularly concerned with two very common types of mottle, *quantum mottle* and *electronic mottle,* which appear in the image, respectively, as *random mottle* or *periodic mottle.*

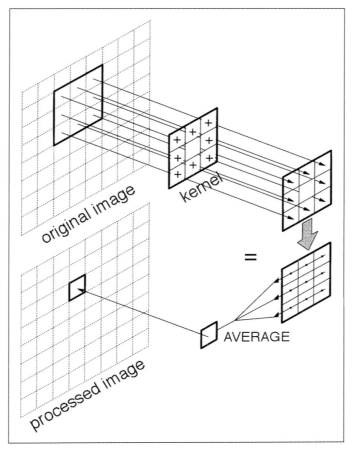

Figure 5-3. Correction for a dead or stuck pixel: A nine-cell kernel first sums the pixel values for the surrounding eight pixels (*top*). These values are then averaged to "fill" the dead or stuck pixel (*bottom*).

Both types of mottle might be considered as "naturally occurring" or unavoidable to some degree. Physicists are acutely aware that all forms of *signals* have an inherent amount of "background noise" in them which consists of small random fluctuations that are not part of the useful signal. In the case of an electrical current passing down a wire, for example, these fluctuations might be caused by natural radiation from space and from the earth passing through the wire, local magnetic fields from other nearby devices, and a host of other causes outside of our control. This becomes especially apparent with very small *microcurrents* such as might be generated in radiation-detection devices. It is a continuing challenge to find ways to "screen out" these small fluctuations to obtain a clean, pure signal.

In the case of medical imaging equipment, the end result of electronic noise in the image itself is *periodic mottle*, that is, mottle which appears with a consistent size and at regular intervals throughout the image, forming a pattern. Some forms of dexel dropout result in whole rows or sets of dexels failing in regular intervals and qualify as periodic in nature.

Quantum mottle occurs from conditions within the *x-ray beam* and follows a natural law called a "Poisson distribution" which applies to all randomly distributed phenomena. You can observe this distribution on a sidewalk any time there is a light rain shower: When very few raindrops have fallen, you can count the number of raindrops in each square of cement on the sidewalk to prove what you observe, that their distribu-

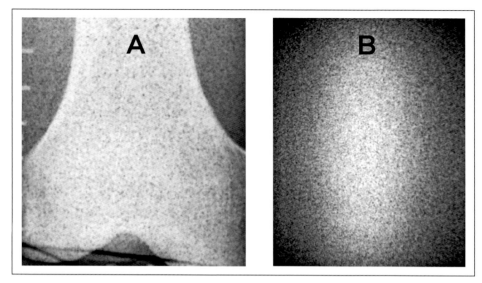

Figure 5-4. Quantum mottle, *A*, results from naturally occurring randomness in the distribution of x-rays within the beam. Electronic mottle, *B*, manifests on a television or LCD monitor screen from small fluctuations in current that occur in all electronic devices.

tion is random, with more drops counted in some squares than others. As a rain shower increases in intensity (increasing the amount of "signal), the cement squares eventually all become saturated, and then it is difficult to tell any more that the actual distribution of raindrops is random. The randomness is still present, but with lots of signal, it is less apparent.

The very same random distribution of x-rays occurs within an x-ray beam. It becomes visible in the image when exposure to the image receptor is very light, such as a light rain shower. In the case of quantum mottle, upon close inspection, *the "speckles" created in the image can be seen to vary in size, and to be randomly distributed throughout the image,* rather than regularly or periodically distributed. When plenty of signal reaches the IR, the random mottle is no longer apparent in the image.

When either periodic or random mottle becomes severe in the image, they may be indistinguishable from each other at the gross observational level, that is, a *severely* mottled image looks about the same whether the root cause was electronic or quantum, as shown in Figure 5-4.

Digital processing can correct for moderate amounts of both random mottle and periodic

mottle. Periodic mottle is best corrected using *frequency filtering algorithms* during *frequency processing,* which will be covered in Chapter 8. These frequency algorithms "attack" a very narrowly-defined *size* of mottle.

For *quantum mottle,* however, it is better to use a *kernel* such as the one shown in Figure 5-3. Kernels can attack a broader range of *sizes* of mottle, cleaning up the various sizes of "speckles" resulting from quantum fluctuations in the x-ray beam.

Field Uniformity

Several flaws are found in the electronics of receptor systems as well as the optical components of a CR reader (such as lenses and optic fibers). Such limitations are inherent in every image acquisition system, and result in an uneven distribution of the "background" density of the resulting images. *Flat field uniformity* is tested on newly manufactured equipment by making a low-level exposure to the entire area of the image receptor without any object in the x-ray beam. The resulting pixel values are compared for the center of the image against each of the four corner areas. Either electronic amplification or computer software can be used to compensate for areas that are outside a narrow range of acceptable deviation from the average, to even out the uniformity of the "flat field" across the IR.

The anode heel effect results in a gradation of x-ray emission from one end of the x-ray tube to the other, causing less exposure at the anode end of the image receptor and more at the cathode end. This can also be compensated to some degree using electronics or software, but since the extent of the anode heel effect is dependent upon the source-to-image distance (SID) used, and SIDs vary, it cannot be fully compensated for.

There are unavoidable variations in the sensitivity of DR detector elements (dexels), and some of the electric charges from the detected exposures must travel down longer wires than others, therefore meeting more electrical resistance. These are just two examples of *electronic response and gain offsets* that must also be corrected for. For CR, there can be some variation in emission from the phosphor plate, and the optic fibers in the reader vary in length and may have optical flaws in them. All of these factors affect the uniformity of the background field for the image, and require correction.

Constructing the Histogram

Before rescaling of the image can take place, the computer must build up a *histogram* of all the image data, in which a *count* is made of all pixels sharing the same pixel value (density, or brightness). This is done for each pixel value possible within the dynamic range of the software (Chapter 2). Visually, the histogram may be thought of as a *bar graph* of the image data as shown in Figure 5-5. For each possible pixel value from "white" to light gray, dark gray and then black, the height of the vertical bar represents how many pixels within the image recorded that value. There is no indication of the *location* of these pixels in the image or what anatomy they represent, just a simple count for each "brightness" value.

For consistency throughout this textbook, we interpret histograms from left-to-right as representing pixel densities from light-to-dark. The histogram software written by some x-ray and photography companies presents the histogram in *reverse,* from dark to light pixel values as one scans from left-to-right. In this case, a "tail" lobe for a type 1 histogram would appear at the *left.*

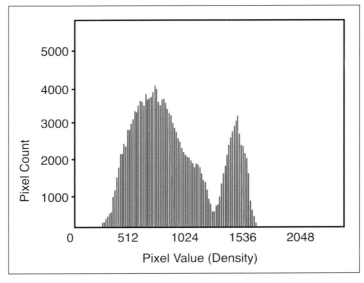

Figure 5-5. The image histogram is actually a *bar graph* depicting the pixel count throughout the image for each pixel value or density. Shown in Figures 5-6 and 5-7, most histograms are depicted as a "best-fit" curve connecting the top of the vertical bars. Densities are usually plotted from left-to-right as light to dark, although this can be reversed.

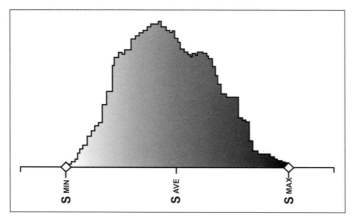

Figure 5-6. Single-lobe histogram typical for a body part covering the entire area of the image receptor, such that there is no "background" density. (From Q. B. Carroll, *Understanding Digital Radiograph Processing* (video), Denton, TX: Digital Imaging Consultants, 2014. Reprinted by permission.)

By identifying just such typical landmarks as the "tail" lobe, one can usually sort out the presentation format of the histogram.

It turns out that histograms acquire generally consistent *shapes* for different body parts, the key distinction being the number of *lobes* or high-points that are generated in the data set. For an image over the abdomen of a large patient that completely covers the IR (image receptor), a single lobe is typical, indicating a few areas of light density (bones), mostly grays (soft tissue), and a few darker areas (gas). This is illustrated in Figure 5-6.

Views of the extremities routinely present significant areas of "raw" x-ray exposure outside of the anatomical part, a "background density" that is pitch black. Chest radiographs can also present a large number of very dark pixels representing both the darkest areas in the aerated lungs and raw exposure above the shoulders and often to the sides of the lower torso. In all these cases, a second lobe may be expected in the histogram to the *right* of the "main lobe" (Figure 5-7). This spike at the right is characteristic for all views that typically present a "raw exposure" background density surrounding the anatomy, and is often referred to as the "tail lobe."

It is possible for a histogram to have *three* lobes: In mammography, the chest wall often presents an unusual number of very light pixels, while the raw exposure outside the breast presents a large number of pitch black pixels (Figure 5-8). Routine radiographs taken under unusual circumstances can also result in a three-lobe histogram, such as an abdomen on a small child (leaving raw background densities to the sides) but also with an area covered by a sheet of lead, a lot of orthopedic hardware (such as scoliosis rods), or when the stomach or colon has a large bolus of barium present (a "solid column" barium procedure as opposed to an air-contrast procedure). In all these cases, the histogram will have a third lobe spiking at the left, indicating an unusually large number of pixels that are very light, as shown in Figure 5-9.

Histogram Analysis

In the processing step called *histogram analysis,* the computer begins by effectively comparing the actual histogram from the exposed image to an *expected histogram shape* for that procedure. If the general shape matches, analysis can proceed without errors. The initial purpose of analyzing the histogram is to *eliminate extreme data that will skew the rescaling of the image, making it come out too dark or too light.*

For example, on views of smaller extremities, the amount of "raw background exposure" surrounding the body part, especially when the

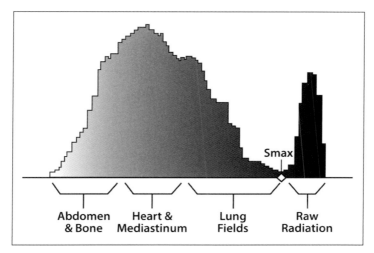

Figure 5-7. For projections that typically present a "raw" exposure or background density surrounding the body part, a second lobe called the *tail lobe* appears in the histogram (*right*). Here, the histogram for a PA chest shows the pixel ranges for chest anatomy, with the tail lobe representing the direct-exposure areas above the shoulders and to the sides of the lower torso. For proper histogram analysis, all pixel values to the right of the S$_{MAX}$ must first be removed from the anatomical data set. (From Q. B. Carroll, *Understanding Digital Radiograph Processing* (video), Denton, TX: Digital Imaging Consultants, 2014. Reprinted by permission.)

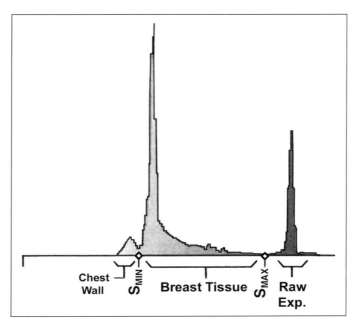

Figure 5-8. The typical histogram for a cranio-caudal mammogram projection includes a third lobe at the left representing very light pixels in the chest wall (see caption to Figure 5-9). (From Q. B. Carroll, *Understanding Digital Radiograph Processing* (video), Denton, TX: Digital Imaging Consultants, 2014. Reprinted by permission.)

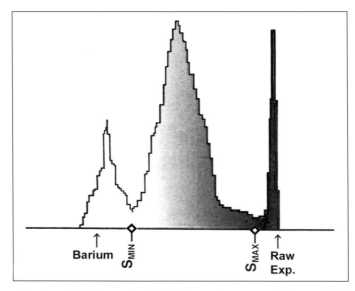

Figure 5-9. A "solid-column" bolus of barium, a lead shield, or a large metallic prosthesis can create a spike in very light densities forming a lobe at the left of the histogram. To prevent errors is histogram analysis, landmarks at both the SMIN and SMAX points must be identified so that excessively light and excessively dark values can be excluded from calculations for digital processing. (From Q. B. Carroll, *Understanding Digital Radiograph Processing* (video), Denton, TX: Digital Imaging Consultants, 2014. Reprinted by permission.)

field is left open, can represent a very substantial amount of the total data acquired. The computer will tend to interpret this as a very dark overall image, and compensate by lightening the image up. However, the more dark background there is, the more the computer will over- correct, such that the bones in the image could be much too light for proper diagnosis. For rescaling to work properly, all or nearly all of the data must represent densities *within the anatomical part,* not "false" densities that are unrelated to the anatomy. Therefore, the spike in the histogram that represents raw background exposure must be identified (a process some manufacturers refer to as *exposure field recognition* or *EFR*), and then this spike must be eliminated from the data set to be operated on.

How can the computer recognize these lobes in the histogram so it "knows" which one to cut out of the data set? Different specific methods may be used by different manufacturers, but our goal here is to give an example of a very basic algorithm that can accomplish this task, using simple subtraction: In Figure 5-10, imagine the computer scanning from *right to left,* subtracting

the pixel counts in each adjacent pair of "bins." (As described earlier, each "bin" actually represents a particular pixel value, density or brightness.) Figure 5-11 is a close-up view of the histogram: In bin #2000, the pixel count is 10 pixels. Moving to the left, bin #1999 holds 20 pixels. Subtracting 20 from 10, we get minus 10. Further to the left, bin #1998 has 35 pixels. Subtracting 35 (bin #1998) from 20 (bin #1999), we get minus 15. *All the resulting subtractions will result in negative numbers until we reach the peak of tail lobe.* Then, as we scan *downward* along the left slope of this lobe, *positive numbers will result from each subtraction.* For example, 30 pixels (bin #1900) minus 25 pixels (bin #1899) is 5, and 25 minus 15 (bin #1898) is 10. When the computer starts scanning up the right slope of the main lobe, the subtractions will result in negative numbers again, (2 – 4 = minus 2, 4 – 5 = minus 1). A simple algorithm can be written instructing the computer to identify the point *SMAX* as the landmark where the results of our subtractions become negative values *for the second time.*

Having thus identified the point *SMAX* where the main lobe ends and the tail lobe begins, we

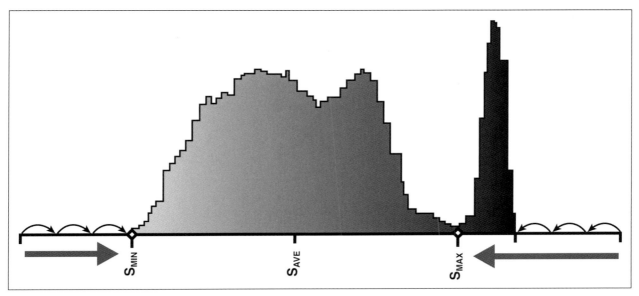

Figure 5-10. The computer scans inward from each end of the histogram, from bin to bin. When a non-zero count of pixels is found, it begins searching for designated landmarks such as SMIN, SAVE, and SMAX. (From Q. B. Carroll, *Understanding Digital Radiograph Processing* (video), Denton, TX: Digital Imaging Consultants, 2014. Reprinted by permission.)

can instruct the computer to remove all information to the right of ***SMAX***, that is, the tail lobe, from the data set before analyzing it. This is just one example of how landmarks in the histogram can be identified.

For histogram analysis, the computer always "scans" into the histogram from the left and from the right, looking for data set "landmarks." One type of landmark is to set a *threshold* to the pixel count that must be exceeded to keep that data for analysis. Most manufacturers use threshold algorithms, for example, to reduce noise. In Figure 5-12, the point labeled ***SMIN*** falls just below the preset threshold count. This pixel count and all pixel counts to the left of it will not be included for analysis. These are very light densities that might represent dead pixels or other forms of noise.

Having eliminated both background densities and noise from the data set, Figure 5-13, the remaining data between points ***SMIN*** and ***SMAX*** all represent *true anatomical structures,* which should result in proper rescaling of the image (discussed in next chapter).

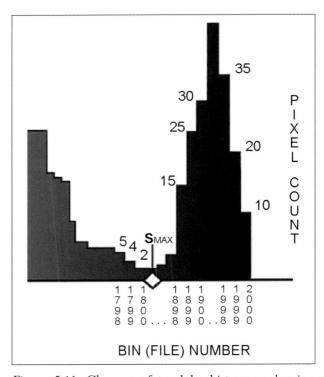

Figure 5-11. Close-up of two-lobe histogram showing pixel counts. Subtracting these from *right-to-left,* upward slopes result in negative differences, downward slopes in positive values.

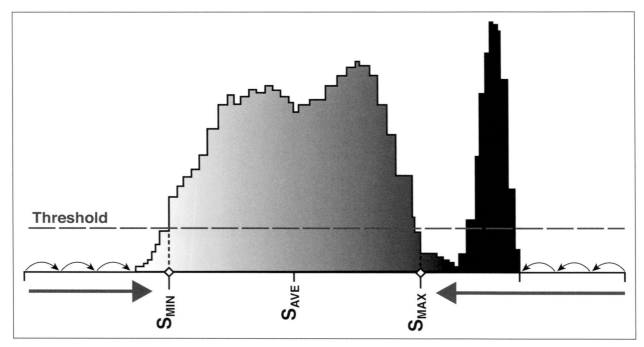

Figure 5-12. Much of the background noise in an image can be removed by *threshold algorithms* that narrow the range, between **S**MIN and **S**MAX, of pixels to be processed. Only pixel values that exceed a minimum pixel count are kept for processing. (From Q. B. Carroll, *Understanding Digital Radiograph* Processing (video), Denton, TX: Digital Imaging Consultants, 2014. Reprinted by permission.)

Types of Histogram Analysis

When the radiographer is at the control console, selecting the *procedure* from the computer menu automatically assigns the type of histogram analysis to be used, based on the shape the acquired histogram is *expected* to have. There are three general types of histogram analysis. To properly identify the landmarks discussed above, *the number of lobes "expected" by the analysis algorithms must match the number of lobes in the actual acquired histogram.*

For example, suppose the subtraction algorithm we used above for the histogram in Figure 5-11 was applied to an actual histogram with only one lobe (as in Figure 5-6). The algorithm was to identify **S**MAX as the point where subtracted values become negative *for the second time.* The computer would complete a search from right-to-left, up and down the single lobe, and then find all zeros for pixel counts rather than a second set of negative subtracted values.

In effect, the computer program "wouldn't know what to do," and might even eliminate the entire image data set, mistaking it for a "tail lobe" since it was the first lobe encountered from right-to-left. With this misinformation, erroneous rescaling would result in a displayed image that would be extremely dark or extremely light.

A practical example of how this might occur is if the procedure were an extremity, but the field was over-collimated into the anatomy on all sides such that there were no "background" densities present in the latent image.

Type 1 histogram analysis is designed to analyze two-lobe histograms with a "tail-spike" representing background densities. *Type 2* analysis is designed for a single-lobe histogram. *Type 3* analysis operates on the assumption that there will be three lobes such as in Figure 5-9, with some metallic material present. Thus, it searches for both an **S**MAX point and an **S**MIN point, so it can exclude both extremes in pixel values from the data set for analysis.

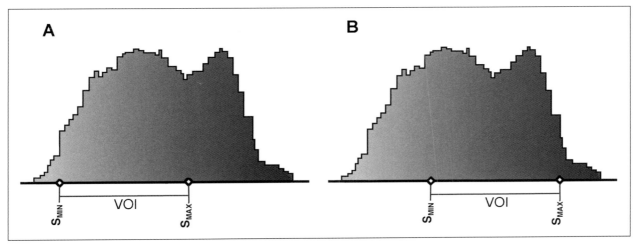

Figure 5-13. With a narrowed range between *SMIN* and *SMAX*, the defined *values of interest (VOI)* can be moved to the left in the histogram to accentuate bony anatomy, *A*, or skewed to the right to accentuate darker soft tissues or the lung fields, *R*.

One way the digital image can be manipulated is by applying algorithms that effectively narrow the range of analyzed data between the *SMAX* and *SMIN* points, and then *skew* the location of this range to the right or to the left within the main lobe of the histogram, as shown in Figure 5-13. This narrowed range is referred to as the *values of interest (VOI)*. Moving the VOI to the left, toward lighter densities, would tend to accentuate bony structures in the image, while moving it to the right would tend to accentuate darker anatomy such as the aerated lung fields. Manufacturers use different proprietary names for features that apply this type of image manipulation at the histogram.

Errors in Histogram Analysis

There are several potential causes for histogram analysis failure. Segmentation errors throw off the histogram. Failure to properly match the type of histogram analysis to the actual acquired histogram leads to errors. Extreme or unusual circumstances may result in a histogram shape that is bizarre and unexpected, such that the analysis algorithms cannot properly identify key landmarks.

An example of such an extreme situation is illustrated in Figure 5-14, a hypothetical histogram for an exposure that was taken of a patient with a large radiopaque prosthesis, but also a corner of a lead apron was draped over the gonads. The presence of a large number of light pixels from the prosthesis, combined with a large number of white pixels from the lead, might result in two spikes at the left of the histogram. Scanning into the histogram from the left, a *type 3* algorithm expects one peak of very light densities and locates SMIN between these two spikes. The very light pixels from the prosthesis are still included in the data analysis, skewing SAVE toward the left. In compensation, the image is rescaled too dark.

Figure 5-15 illustrates how histogram analysis might fail from "pre-fogging" of a CR plate from leaving it out in the x-ray room for a period of time. CR plates are extremely sensitive to scatter and background radiation *accumulated during storage.* If the plate acquired a lot of exposure prior to use, the very lightest, nearly white densities would be eliminated from the acquired histogram, shown from *A* to *B* in Figure 5-15. (There might also be a spike of light gray densities as shown in *B*.) If the SMIN point is shifted to the right as shown in *B*, SAVE will also be skewed to the right, and computer compensation may result in a light image with low contrast.

Digital processing algorithms have become increasingly robust, and it is difficult to "throw

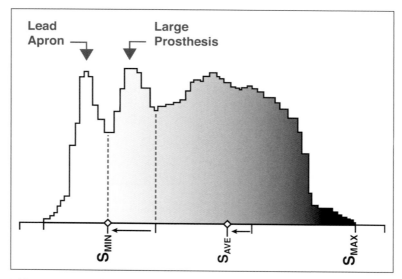

Figure 5-14. A potential histogram with *two* left lobes when both a large prosthesis and a lead apron were present in the exposure field. Scanning from the left, the computer identifies S$_{MIN}$ between these two lobes, (compare to Figure 5-9), pulling S$_{AVE}$ to the left and causing improper rescaling. (From Q. B. Carroll, *The Causes of Digital Errors* (video), Denton, TX: Digital Imaging Consultants, 2014. Reprinted by permission.)

off the computer." For histogram analysis errors to occur, it requires either a very unusual set of conditions such as just described above, or a very extreme condition such as very large amounts of scatter radiation. Observe the **S*MAX*** point in Figure 5-16 for example: A certain amount of scatter radiation produced during exposure might raise the low point before the tail spike upward, yet it might not exceed the threshold level. In this case, histogram analysis will *not* be corrupted and the image will still be processed correctly. *Digital processing is generally very adept at correcting for moderate amounts of scatter radiation caused during exposure.*

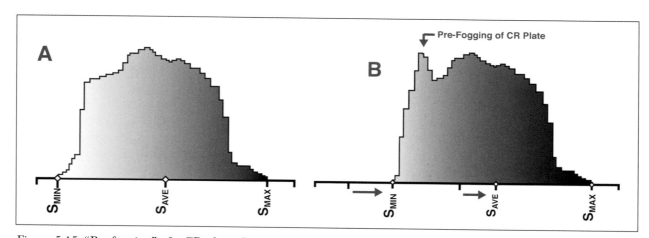

Figure 5-15. "Pre-fogging" of a CR plate eliminates the lightest densities (to the left in *A*), shifting SMIN to the right as shown in *B*. Since this also will pull the average pixel value (S$_{AVE}$) to the right, improper rescaling may result. (From Q. B. Carroll, *The Causes of Digital Errors* (video), Denton, TX: Digital Imaging Consultants, 2014. Reprinted by permission.)

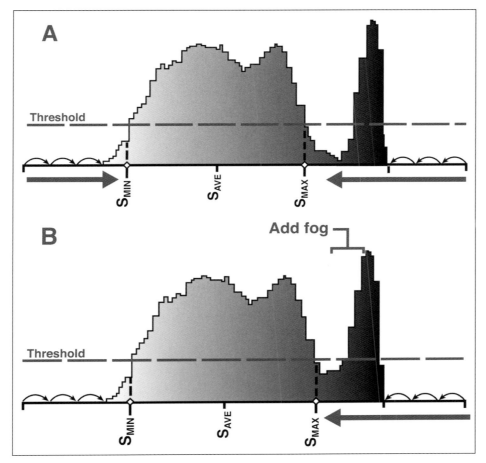

Figure 5-16. Histogram analysis is unimpeded by most instances of scatter "fogging" *during exposure.* In histogram *B*, the addition of an area of dark "fog" thickens the tail spike and raises the point *S$_{MAX}$*. However, since *S$_{MAX}$* still falls *below the threshold* line set for histogram analysis, no errors result in the identification of histogram landmarks, and the image is correctly processed.

As discussed in the next chapter, both kernels and frequency processing can largely eliminate specific fog patterns that are common to certain projections such as the lateral lumbar spine. There are many ways in which the effects of scatter are now compensated for, and only the most extreme and unusual conditions result in an image that is of unacceptable diagnostic quality, requiring a repeat exposure to be taken.

With histogram analysis now completed, we are ready to move on to the first major processing step for the digital radiograph: *Rescaling* the image.

Chapter Review Questions

1. Unlike the *four* steps of film processing, the _____ steps of digital radiograph processing can be applied in different order or even repeated.
2. *Preprocessing* is defined as all computer operations designed to compensate for flaws in image _____.
3. *Postprocessing* makes refinements to the image customized to a particular anatomical _____.
4. The digital processing step most closing associated with the concept of simple *processing* of the image is _____.

5. It is only during rescaling that the acquired image takes on the level of _____ necessary for diagnosis to take place.

6. Failure of a CR system to separate two or more individual exposures taken on the same plate is referred to as _____ failure.

7. Dexel dropout is typically corrected using _____ -reduction software.

8. A *kernel* can interpolate a pixel value to replace a "dead" pixel by averaging the values of the _____ pixels surrounding the dead pixel.

9. Whereas quantum noise in the x-ray beam results in *random* mottle, *electronic* noise typically results in _____ mottle in the image.

10. Whereas frequency-filtering algorithms are best used to correct for periodic mottle of a consistent size, _____ are better at correcting the broader size range of quantum mottle.

11. The anode heel effect, variations in electronic response and gain offsets all contribute to the necessity of correcting flat-_____ uniformity as part of preprocessing

12. The initial histogram is actually a _____ graph of simple pixel counts for each pixel value.

13. If a pitch-black background density and a massive light area (such as the chest wall in a mammogram) are both present, the acquired histogram for a radiographic image can have _____ lobes or high-points.

14. For rescaling to work properly, nearly all the data it uses must represent densities within the _____, rather than extreme data such as from background exposure or lead aprons.

15. *Exposure field recognition* algorithms identify any "raw" _____ exposure area in the image.

16. To reduce noise, most manufacturers use _____ algorithms for histogram analysis.

17. The number of lobes "expected" by the type of histogram analysis used must *match* the number of lobes in the actual acquired histogram in order to properly identify key _____ in the histogram.

18. The displayed image can be manipulated by narrowing the *range* of analyzed data, called the _____ of _____, and locating this range in a targeted portion of the histogram.

19. Only the most _____ or _____ circumstances that result in a bizarre shape to the histogram can cause errors in histogram analysis.

20. Although histogram analysis is robust against most scatter "fogging" events caused during *exposure,* it is more likely to fail from _____ of a CR cassette during storage prior to use.

Chapter 6

RESCALING (PROCESSING) THE DIGITAL RADIOGRAPH

■ ■ ■ ■ ■ ■ ■ ■ ■ ■ ■ ■ ■ ■ ■ ■ ■ ■ ■

Objectives

Upon completion of this chapter, you should be able to:

1. State the simplest mathematical operations for changing the brightness and contrast of the incoming latent image.
2. Describe how these operations affect the position and shape of the image histogram.
3. Compare the diagnostic qualities of the incoming "raw" latent image to those of an image *normalized* by rescaling.
4. Explain the basic mathematical process of rescaling.
5. Define *S values, Q values,* and *remapping.*
6. State the limitations of what rescaling can achieve in the final image.

Using mathematical formulas and functions, a set of numerical data can be manipulated in all sorts of wondrous ways. This allows us to increase or decrease the overall darkness of the image (*levelling,* or changing the window level), and to increase or decrease the contrast of the image (*windowing,* or adjusting the window width).

Anything which alters the *shape* of the histogram represents a change in image contrast or gray scale. There are a number of ways that this can be done mathematically, and we'll give two of the simplest examples here: Using exponential functions and numerical rounding.

Table 6-1 lists two simplified sets of pixel values, one for a standard image and one for a high-contrast image. Note that the average pixel value for both images, the mid-point *M*, is identical, set at an output of 128. This means that the window level, or overall brightness, has not been changed, only the contrast. In column A, the pixel values increase and decrease in increments of 2. In column B, for the high-contrast image, the pixel values change in increments of 4. This is a greater ratio of *difference* between each pixel value and the next, and thus represents high contrast as we defined it in Chapter 4. This is the type of result we get when exponential formulas are applied to the data set.

A simplified example (in English) of a computer algorithm using exponential relationships might read as follows: "Begin at the median value *M*, the average pixel value for the image. For the next darker value, add 2. For the next, add 4. For the next, add 8, and so on. For the next lighter (lesser) value, subtract 2. For the next, subtract 4. For the next, subtract 8, and so on. The result will be a higher contrast image.

Figure 6-1 illustrates how numerical rounding can change contrast or gray scale. As described in Chapter 2, all digital information has already been rounded—this is effectively what changes it from analog data into digital or discrete data. For histogram *A* in Figure 6-1, pixel values are rounded to the nearest one-tenth. The pixel value for bin #800 is 3.1, for bin #801 it is 3.2, #802 is 3.3, and #804 is 3.4. What happens if we

Table 6-1
ACTUAL LOOK-UP TABLE FORMAT:
MEDIUM AND HIGH CONTRAST FOR SAME INPUT

Table A Medium Contrast			Table B High Contrast	
INPUT	OUTPUT		INPUT	OUTPUT
30	156		30	240
28	152		28	224
26	148		26	208
24	144		24	192
22	140		22	176
20	136		20	160
18	132		18	144
16	128	**M**	16	128
14	126		14	112
12	122		12	96
10	118		10	80
8	114		8	74
6	110		6	58
4	106		4	42
2	102		2	26
0	98		0	10

apply a "harsher" rounding process, such that each pixel value is rounded to the nearest *one-fifth* or 0.2? The pixels in bin #800 will be rounded up from 3.1 to 3.2 and join those pixels in bin #801. The pixels in bin #802 will be rounded up from 3.3 to 3.4, and join the pixels in bin #803. The resulting histogram *B* shows that when the difference between each bin is increased to 0.2 (increasing contrast), *there are fewer vertical bars in the graph, representing fewer pixel values present in the image.* This is the definition of *shortened gray scale*—fewer densities, a lesser range of densities, in the image.

To what extent can we alter the position and shape of the image histogram mathematically? Shown in Figure 6-2, from *A* to *B*, lightening the image is tantamount to shifting the histogram to the *left,* and involves the simplest of computer algorithms, e.g., "Subtract 0.3 from all pixel values." The instruction, "Add 0.4 from all pixel values" would shift the histogram to the *right* and result in a darker image overall. In clinical practice, this is *leveling* or changing the window level. These operations shift the *center-point (SAVE in the previous chapter)* of the histogram to the left or right, but never alter the shape of the histogram.

Figure 6-2, *B* to *C* demonstrates how changes in gray scale or contrast can alter the side-to-side *range* of the graphed histogram. This is the effect of the rounding adjustment from Figure 6-1, and in clinical practice is called *windowing* or chang-

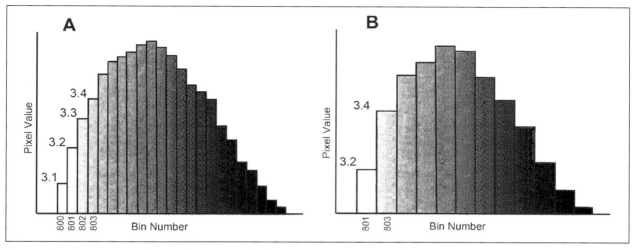

Figure 6-1. One way to shorten the gray scale (narrow the window width) is by simply increasing the *severity* with which pixel values are rounded during digitization. Histogram *A* results from rounding pixel values to the nearest *tenth (0.1),* histogram *B* by rounding to the nearest *fifth (0.2).* Note that for *B*, there are fewer vertical bars (pixel values available). This is shortened gray scale.

ing the window width. To summarize, on a histogram graph, leveling changes the side-to-side *position* of the histogram (measured at its median or middle point), and windowing changes the side-to-side *range* of the histogram.

However, shown in Figure 6-2 *C* to *D*, there is one aspect of the histogram that cannot be changed, and that is the original *height* of each bar. This height represents the original pixel count which was made by effectively scanning across the image and locating each pixel that held a particular value to count it. We cannot change *which* pixels contained which original values. In other words, we can play all day with pixel values, but not with pixel counts.

"Normalizing" the Image

Normalizing is the essence of rescaling. Both the overall brightness of the image and its degree of gray scale or contrast are manipulated mathematically until it takes on the "normal" appearance of a conventional radiograph. If a "raw" digital image were taken directly from a DR detector plate or from a CR phosphor plate and displayed on a monitor, it would have extremely poor quality, especially in terms of image contrast. In Chapter 5, Figure 5-1 demonstrates the typical appearance of such a "raw" digital image from the image receptor before any processing has occurred. This is a ghost-like image with almost no contrast, and certainly inadequate for diagnosis.

When compared to conventional film imaging, we have identified the primary advantage of digital imaging as its ability to enhance subject contrast in the image. This enhancement occurs only upon actual digital rescaling by the computer—it is yet not present in an image taken directly from the image receptor system, whether CR or DR. It is the rescaled image that will now be subjected to additional default *post-processing* in order to refine and customize it according to the specific anatomical procedure and preferences of the diagnostician before it is initially displayed on a monitor.

Because of rescaling, the final displayed digital image nearly always has an ideal level of brightness and balanced gray scale, *regardless of the specific radiographic technique used upon initial exposure.* Conventional film-based radiographs were extremely sensitive to the radiographic technique used: They often turned out too dark when excessive technique was used or too light

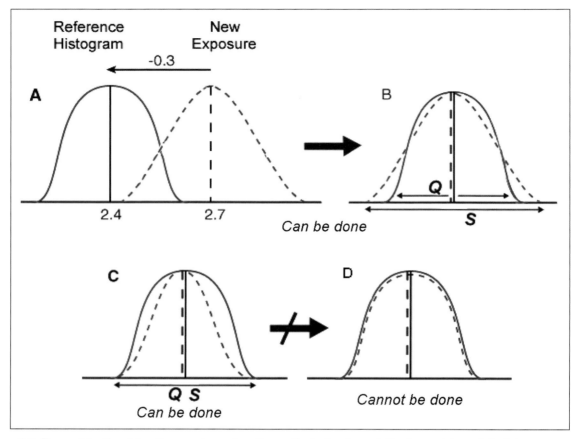

Figure 6-2. From *A* to *B*, subtracting a set number from all pixel values shifts the position of the histogram, resulting in a lighter image. From *B* to *C*, the range of pixel values is narrowed by an operation like the one illustrated in Figure 6-1. The *vertical shape* of the histogram *cannot be changed* (*C* to *D*) because this would involve altering the original pixel counts at each value.

when insufficient technique was used. As shown in Figure 6-3, this is no longer the case. The primary role of radiographic technique now is simply to produce sufficient signal reaching the IR for the computer to successfully process. Digital processing by the computer is so robust in making corrections that the final displayed image is nearly always sufficient for diagnosis. Corrections made by the computer fail only under the most unusual circumstances or extreme changes.

Q Values and the "Permanent" LUT

Rescaling can be achieved either electronically or with software, but software programming has several advantages and is used for most applications. We must manipulate the measured pixel values shown along the bottom of the histogram in Figure 6-4, which we now designate as S values. The key that allows this manipulation is to *algebraically assign the same labels to the incoming data regardless of what the data actually is.* We will designate these new, standardized labels as *Q values* which are stored in what we might call a "permanent" look up table, or LUT.

Table 6-2 provides an example of a permanent LUT where 1024 Q values are assigned preset pixel values that represent certain brightness levels for pixels in the image. In this case, the value *Q minimum (QMIN)* is permanently assigned a pixel value (gray shade) of 511 *for dis-*

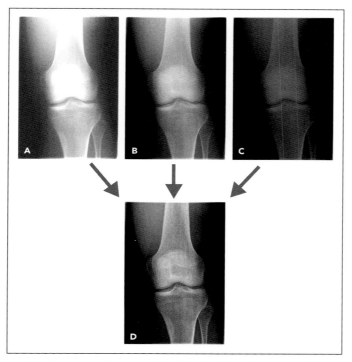

Figure 6-3. Rescaling can nearly always produce an ideal level of brightness and balanced gray scale, *regardless of the specific radiographic technique used upon initial exposure,* only provided that the exposure technique delivered sufficient information at the image receptor. (From Q. B. Carroll, *Radiography in the Digital Age,* 3rd ed. Springfield, IL: Charles C Thomas, Publisher, Ltd., 2018. Reprinted by permission.)

play on a monitor. The value $Q2$ will always be displayed as a pixel value of 512, $Q3$ as 513, and so on up to Q *maximum (QMAX),* which will always be displayed as a pixel value of 1534. If the dynamic range of the digital processing system is from 1 to 4096, we can see that there is room left for a displayed image whose actual gray scale ranges from 511 to 1023 to be "windowed" up and down by a radiographer after it is initially displayed.

As described in the previous chapter, in the histogram for the incoming image from the IR, each measurement taken from a DR dexel or CR pixel is designated algebraically as an S value. We might think of each *S value* along the bottom of the histogram as a "bin" into which pixels may be temporarily placed. Each bin is actually a separate computer file. (Remember that the computer keeps a map of where these pixels actually belong within the spatial matrix, but for histogram analysis, we place all pixels

sharing a particular measured value in the same "bin." For example, the pixels A-3, D-1, F-8, and K-13 may have all shared the same measured pixel value of 2012, but they are all placed in the "bin" labeled *S5.*

We now simply write a computer program that reassigns these incoming S values as Q values in the permanent LUT as shown in Table 6-3. Whatever the original pixel value for *S5* was, it will now be reset to *Q5* from the permanent LUT. Referring to Table 6-2 for our example above, the actual pixel value will be changed from the original 2012 to 515. Pixels A-3, D-1, F-8 and K-13 will all now have a pixel value of 515 when the image is actually displayed. *Regardless of what the incoming pixel values were, the output pixel values are always adjusted to the same output Q values set by the permanent LUT.* This is illustrated graphically in Figure 6-5. The effect is that an input image that is too dark or too light will always be outputted and displayed at a

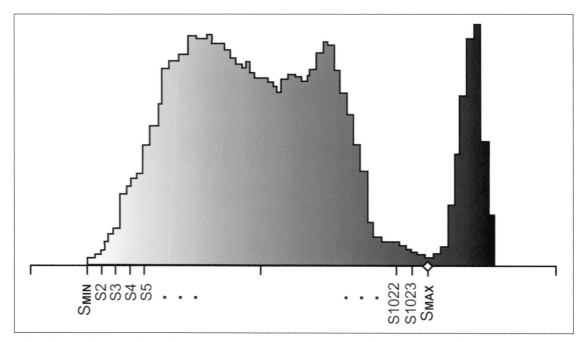

Figure 6-4. For rescaling, each pixel count is treated as a separate "bin" (or computer file) of data. Each bin is given an algebraic label called the *S value* (SMIN, S2, S3 and so on up to SMAX). These labels are used in Table 6-3 to "remap" the image. (From Q. B. Carroll, *Radiography in the Digital Age,* 3rd ed. Springfield, IL: Charles C Thomas, Publisher, Ltd., 2018. Reprinted by permission.)

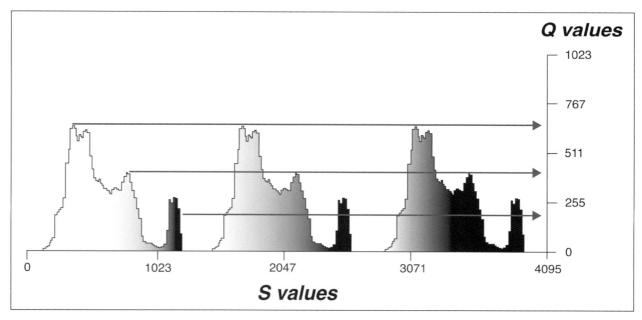

Figure 6-5. No matter where the histogram of the original latent image lies, it can always be *remapped* to *Q* values of the "permanent" LUT (right) by rescaling. This nearly always results in an "ideal" image (Figure 6 -3). The new *Q* values are the "for processing" values that will be used for gradation processing (next chapter). (From Q. B. Carroll, *Radiography in the Digital Age,* 3rd ed. Springfield, IL: Charles C Thomas, Publisher, Ltd., 2018. Reprinted by permission.)

medium level as shown in Figure 6-3. This process is often referred to as *remapping* the pixel values.

For all this to work, the *range* of S values must exactly match the range of Q values in the permanent LUT, so the incoming histogram can be effectively "lined up" with the histogram for the permanent LUT, and each "S bin" will match to a corresponding "Q bin." Regular digitization of the incoming image from the IR facilitates this matching process. Referring again to Figure 6-1, we see that the incoming set of pixel values can be rounded up or down by the ADC with different levels of "severity." When they are severely rounded (**B** in Figure 6-1), fewer S values result. In this way, we can adjust the range of incoming S values to match the range of Q values in the permanent LUT.

Rescaling has the power to align image brightness levels perfectly, but can only align overall image gray scale *partially* by aligning the maximum and minimum Q values. To further adjust image gray scale or contrast, *gradation processing* is required as described in the next chapter. For gradation processing, the rescaled data set is fed into a procedure-specific LUT for fine-tuning based on the anatomy to be demonstrated. This *anatomical LUT* is automatically set when the radiographer selects a radiographic procedure from the menu on the x-ray machine console.

Physicists' Nomenclature

Physicists have now adopted the terms, *Q values* for pixel values that have not yet been rescaled, *Q_K values* for values that have been rescaled but not yet gradation processed, and *Q_P values* for pixel values "ready for presentation" that have completed all processing operations and form the initially displayed image on the monitor. For the student, we will simplify these designations throughout this book by following the practice of several manufacturers of using *S values* when referring to incoming data that has not been rescaled, and *Q values* for rescaled (processed) data.

Table 6-2
Q VALUES STORED
IN THE PERMANENT LUT

Q Values Stored in the Permanent LUT	
Q_{MIN}	511
Q2	512
Q3	513
Q4	514
Q5	515
–	–
–	–
Q1022	1532
Q1023	1533
Q_{MAX}	1534

Adapted from *Understanding Digital Radiograph Processing,* Midland, TX: Digital Imaging Consultants, 2013. Reprinted by permission.

Table 6-3
REASSIGNING S VALUES AS Q VALUES

Reassigning S Values as Q Values		
Algorithm:		
Set S_{MIN}	=	Q_{MIN}
Set S2	=	Q2
Set S3	=	Q3
Set S4	=	Q4
Set S5	=	Q5
–		
–		
Set S1022	=	Q1022
Set S1023	=	Q1023
Set S_{MAX}	=	Q_{MAX}

Adapted from *Understanding Digital Radiograph Processing,* Midland, TX: Digital Imaging Consultants, 2013. Reprinted by permission.

Chapter Review Questions

1. Changes made to the contrast or gray scale of the image will alter the _____ of the histogram

2. Mathematically, we can increase the contrast of an image by simply _____ the increments between output pixel values.

3. A "harsher" or more sever rounding process at the ADC is one way to shorten image _____ _____.

4. Subtracting the same number from all pixel values in the image will shift the histogram to the _____.

5. Changing the window level alters the side-to-side position of the histogram, but changing window *width* alters the side-to-side _____ of the histogram

6. The one thing we *cannot* change in the histogram is the original _____ of each bar, which represents a pixel _____.

7. "Normalizing" the appearance of the image to a diagnostic level of contrast is the essence of _____ the image.

8. Rescaling is able to nearly always achieve ideal levels of brightness and gray scale regardless of the particular radiographic _____ used for the initial exposure.

9. The key to rescaling is to algebraically assign the same _____ to incoming data.

10. A "permanent LUT" stores preset pixel values for the output image called _____ values.

11. Each actual measurement taken from the DR dexel or CR pixel for the incoming latent image is designated algebraically as a(n) _____ value.

12. Regardless of what the incoming pixel values of the latent image are, they are always adjusted to the same output Q *values* set by the permanent _____.

13. For rescaling to work, the *range* of incoming S values must be rounded to _____ the range of preset Q values.

14. Rescaling can completely adjust image brightness, but can only _____ adjust image gray scale. This is one reason why gradation processing follows.

Chapter 7

DEFAULT POSTPROCESSING I: GRADATION PROCESSING

Objectives

Upon completion of this chapter, you should be able to:

1. Define and differentiate between the three domains in which digital images can be processed. Give examples of each.
2. Explain how a *kernel* can be used to alter a digital image.
3. Differentiate between point-processing, local processing, and global processing operations.
4. Interpret gradient curves as they relate to the brightness and contrast of an image.
5. Define gradation processing, function curves, and intensity transformations.
6. Describe how anatomical look-up tables (LUTs) are used in gradation processing.
7. Relate leveling and windowing to gradation processing.
8. State the causes and implications of data clipping.
9. Explain how *dynamic range compression (DRC)* is used both to save computer storage and to achieve tissue equalization in the displayed image.

Digital Processing Domains

Imagine you are in the middle of a car dealer's lot. As you look around yourself, there are *three* different ways you can broadly classify the various cars you can see. First, you could sort them by their spatial location: *There is one car nearby to my left, there is a car directly in front of me but much farther away, and there is a car midway to my right.* The second way you could classify them is to group them by color or by shade: *I see six very dark cars, two very light cars, and three mid-shaded cars.* Finally, you might choose to group them by size: *There are three very large cars, five mid-size cars, and three compact cars.*

These are everyday examples of the three general approaches for digitally processing any image. Shown in Figure 7-1, before pixels or objects within the field of view can be acted upon by the computer, we must first choose which of the three methods of *sorting* them we are going to use: Shall we sort them by *location*, by *intensity (shade)*, or by *size*? Depending on which method we choose, all kinds of different operations can be executed to change and manipulate the image. These differ so much that we refer to the three general methods of digital image processing as *domains:*

The *spatial* domain

The *intensity* domain

The *frequency* domain

In the spatial domain, pixels are acted upon according to their *location* within the image matrix. In the intensity domain, pixels are oper-

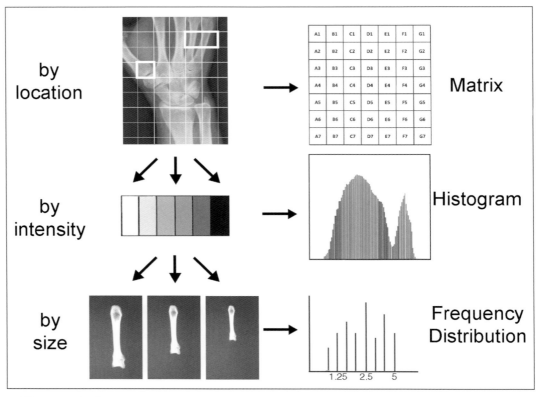

Figure 7-1. Three methods to sort any image for digital processing: Sorting the image by spatial location results in a matrix. Sorting the image by pixel intensity results in a histogram. Sorting the image according to the size of the objects or structures in it results in a frequency distribution.

ated upon based on their *pixel values,* that is, how bright or how dark they are. A brighter pixel is reflecting or emitting more *intense* light. The computer formulas used to make them darker (increase "density") or lighter are referred to as *intensity transformation formulas,* a fitting term since their darkness or brightness, degree of light emission, or even the amount of electricity flowing through them, can all be expressed in terms of how *intense* the electricity, light, or brightness is.

The frequency domain is unique in that it addresses *objects* in the image rather than *pixels.* This will be explained in detail later, but here is the basic concept: A very large object such as a bowling ball, if dropped into a swimming pool, will make big (long) waves. A very small object of similar density, such as a marble, dropped into the same pool of water, will make very small (short) waves. Large objects are associated with long wavelengths, and small objects are associ-

ated with short wavelengths. We know from physics that for waves which travel at a constant speed, their wavelengths are *inverse* to their frequencies. For example, for long waves such as radio waves, only a few waves will strike you per millisecond, whereas for short waves like x-rays, many thousands can strike you each millisecond. Long waves have low frequencies, short waves have high frequencies.

Now, we can measure the *size of an object* within the radiographic image by taking a single row of pixels and counting how many of those pixels are occupied by the one object in question. Larger objects will occupy more pixels along that row, small details will occupy only a few pixels. Large objects are considered as *low-frequency* objects, because they are associated with long waves. Small details are considered as *high-frequency* objects, because they are associated with very short waves. We can thus sort out objects within the image according to their size, execute computer operations on them while they are in the frequency domain, and then re-insert these objects back into spatial matrix where the computer has logged their locations (placing them back into the spatial domain).

Likewise, when we execute intensity operations on an image, the pixels must first be moved from the spatial domain (the matrix) into the intensity domain (the histogram) where they are sorted into separate files according to their pixel values. After intensity operations are completed, they are placed back into the spatial matrix to form the final displayed image. *The digital image always begins and always ends in the spatial domain.*

Shown in Figure 7-1, when an image is sorted by the spatial location of its pixels, the result is the familiar image matrix. When an image is moved into the intensity domain and sorted according to the pixel values present, the result is a histogram. When an image is moved into the frequency domain and is sorted accorded to the size of the structures throughout the image, the result is a frequency distribution.

The frequency distribution plots the number of objects against their size, that is, each vertical bar in the graph represents the number of objects in the image of a particular size. Just like a histogram, each vertical bar also represents a separate computer file. To make a change to all of the objects of a particular size (such as only the fine details) without affecting anything else in the image, we simply apply a mathematical formula to everything within that selected computer file. The frequency distribution graph looks much like a histogram, only with fewer vertical bars—a few hundred rather than a few thousand. The frequency distribution graph in Figure 7-1 is somewhat simplified to distinguish it from the histogram.

Spatial Domain Operations

Examples of spatial domain operations include magnification or "zooming", translation ("flipping" the image right for left), inversion ("flipping" the image top for bottom), image subtraction, and all kernel operations.

A *kernel* is a submatrix that is passed over the larger image matrix executing some mathematical function on each pixel. (This is similar to a spreadsheet computer program which can apply formulas to cells of data contained within a table.) For the 3-by-3-cell kernel shown in Figure 7-2, as the kernel moves from left-to-right one column at a time, a particular pixel that falls within the bottom row of the kernel will first have 3 subtracted from its pixel value, then its value will be multiplied by 2, then it will be added to 4. When the kernel finishes a particular row, it indexes down one row and again sweeps left-to-right all the way across the next row. The lower kernel in the diagram illustrates that all nine cells will eventually sweep over each pixel centered within the kernel. After all rows in the image are completed, the kernel is passed over the entire image again *vertically,* sweeping top-to-bottom for each column. In this way, every pixel in the image is subjected to the effect of every cell in the kernel both horizontally and vertically. By using different values or formulas in the cells of a kernel, all kinds of changes can be made to the image.

Spatial domain operations can be further subdivided into three categories: 1) point processing operations, 2) area processing operations, and 3) global operations. *Point processing* operations perform a specific algorithm on each individual pixel in sequence, pixel by pixel or "point by point." In image subtraction, for example, the value contained in each specific pixel is subtracted from the value contained in the corresponding pixel from another image, but these pixels are identified by their *location* in the spatial matrix, not by the values they contain.

Area or "local" processing operations execute a mathematical function on a subsection of the image, only a designated local group of pixels. For example, a CT technologist can use a cursor on the monitor screen to demarcate a portion of the image she wishes to magnify. At the push of a button, this area will then be "zoomed" up to fill the display screen. To do this, the value for each single pixel will be spread out across an area of 4 hardware pixels on the monitor, or with more magnification, 9 hardware pixels.

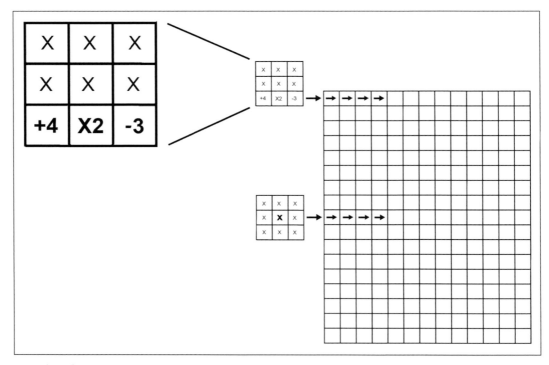

Figure 7-2. A *kernel* is a smaller "submatrix" that passes over the larger matrix of the image executing some mathematical function. Here, for each image pixel, the lower row of the kernel will first subtract the number 3 from the original pixel value, then multiply that result by 2, then add 4 as the kernel passes from left to right. Shown below, pixel values above and below are also being altered by the functions in the first and third row of the kernel. After all rows are swept over, the kernel passes *vertically* over the entire image again, column by column.

Each displayed image pixel occupies multiple hardware monitor pixels.

Global processing applies some massive operation across the entire image. All image reorientations, such as rotating the image, inverting the image, or translating the image (flipping it left for right), are global operations. Figure 7-3 shows how image translation is actually a fairly simple spatial operation; translation leaves only the very middle column intact. The columns immediately to the right and left of this middle column exchange pixel values. The second pair of columns to either side exchange pixel values; the third pair of columns exchange pixel values, and so on to the farthest right and left columns of the image. In Figure 7-3, column *E* is the middle column. If one examines the single row of pixels, **row 6**, across the image, we see that translation exchanges pixel D6 for F6, C6 for G6, B6 for H6, and then A6 for I6. Inverting the

image applies the same simple exchanges, only vertically between rows instead of horizontally between columns.

Intensity Domain Operations

All gradation processing, the focus of this chapter, is carried out in the intensity domain. After the initial image is displayed, any "levelling" and "windowing" by the operator is effectively a readjustment to the gradation processing, and so windowing is usually an intensity domain operation, (although it can also be done in the spatial domain using kernels). Also included in the intensity domain are construction of the original histogram for an image, and histogram analysis, covered in Chapter 5. From the histogram, we see that pixels from various locations are grouped together in "bins" or computer files that share the same pixel value. When a

A1	B1	C1	D1	E1	F1	G1	H1	I1
A2	B2	C2	D2	E2	F2	G2	H2	I2
A3	B3	C3	D3	E3	F3	G3	H3	I3
A4	B4	C4	D4	E4	F4	G4	H4	I4
A5	B5	C5	D5	E5	F5	G5	H5	I5
A6	B6	C6	D6	E6	F6	G6	H6	I6
A7	B7	C7	D7	E7	F7	G7	H7	I7
A8	B8	C8	D8	E8	F8	G8	H8	I8
A9	B9	C9	D9	E9	F9	G9	H9	I9
A10	B10	C10	D10	E10	F10	G10	H10	I10
A11	B11	C11	D11	E11	F11	G11	H11	I11
A12	B12	C12	D12	E12	F12	G12	H12	I12
A13	B13	C13	D13	E13	F13	G13	H13	I13
A14	B14	C14	D14	E14	F14	G14	J14	I14
A15	B15	C15	D15	E15	F15	G15	J15	I15

Figure 7-3. *Translation* "flips" the image left for right by exchanging the pixel values between cells D6 and F6, then those between C6 and G6, B6 and H6, then A6 and I6 and so on progressively away from the middle column. Image translation is a *global* operation. (From Q. B. Carroll, *Radiography in the Digital Age,* 3rd ed. Springfield, IL: Charles C Thomas, Publisher, Ltd., 2018. Reprinted by permission.)

mathematical operation is then applied to any selected computer file, it affects all pixels throughout the image holding that particular original value, *regardless of the locations of those pixels* in the matrix.

Frequency Domain Operations

Examples of operations in the frequency domain include *smoothing, edge-enhancement* and *background suppression.* These operations are executed on *structures* or *objects* within the image rather than on pixels. The objects are identified, sorted and grouped by their *size* into separate computer "bins" or files. Several different operations, generally referred to as *detail processing operations,* can be applied to any selected file. By targeting a specific *size* of structures to affect in the image, the results can seem almost magical, such as being able to enhance the *local contrast of*

only fine details, without changing the overall contrast appearance of the image as a whole (from a distance, for example). The frequency domain is not as intuitive for the student to understand, and will be fully explained in the next chapter.

Gradation Processing

In the list of postprocessing operations presented in Table 5-1, Chapter 5, step #5 is gradation processing. The purpose of *gradation* or *gradient* processing is to tailor the final image brightness and contrast according to the anatomy and predominant pathologies to be displayed, customizing them to the procedure.

Gradation is defined as a gradual passing from one tint or shade to another. In radiography, this relates to the *gray scale* of the image and how we can manipulate it. The *gradient curves* graphed in Figure 7-4 were widely used with film radiogra-

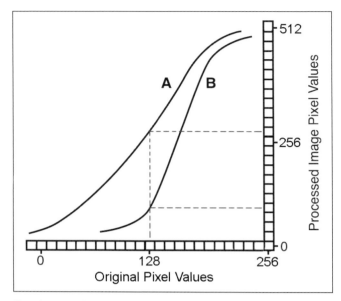

Figure 7-4. Gradient curve **B** indicates the higher contrast output image by a steeper slope to its "body" portion. Curve **A** represents a darker overall output image by its position further to the left. (From Q. B. Carroll, *Radiography in the Digital Age,* 3rd ed. Springfield, IL: Charles C Thomas, Publisher, Ltd., 2018. Reprinted by permission.)

phy and have been used for decades to describe both the brightness (density) of an x-ray image and its gray scale or contrast, especially as it relates to the image receptor. Some manufacturers display a gray scale curve superimposed over the histogram for each image. Each of these curves has a "toe" and a "shoulder" portion where the curvature is greater, but these represent only the greatest extremes in exposure to x-rays. Here, we are concerned primarily with the (nearly) "straight-line" portion or "body" of each curve, because it represents a range of exposures that are typically used in the daily practice of radiography.

The body of curve **A** is different from that of curve **B** in two ways; First, note that curve **A** has a shallower slope. We would describe curve **B** as steeper. For curve **B**, as we read along the bottom of the graph we see that as exposure is increased left-to-right, this image shows a more dramatic change in the pixel values displayed (quantified at the right of the graph). Visually, curve **B** represents a high contrast image, whereas curve **A** is an image with longer gray scale or a more gradual progression from light to dark densities.

In the dictionary, a *gradient* is defined as a "part sloping upward or downward," measuring the rate of graded ascent or descent. In Figure 7-4, then, a high gradient represents high contrast and a low gradient low contrast. Because of this relationship, gradation processing is also often referred to as *gradient processing.*

One more look at Figure 7-4 reveals that gradient curves also represent the overall brightness or density of a radiographic image. Can you describe which image is darker overall? To clarify, choose a specific level of exposure along the bottom such as the original pixel value 128, extend a vertical line upward from this point, and determine which curve is higher along the scale of pixel values, indicating a darker image. We generally look at the mid-portion of each curve to make this determination, and here you can see that normally (though not always) the "higher" of the two curves is also the farthest to the left in the graph.

Thus, gradient curve graphs tell use both the average brightness level (or density) of an image by its left-to-right position, and its contrast level by how steep the slope of the curve is. Note that at the display monitor, these are precisely the two aspects of the final image that we commonly change by "windowing." The window *level* controls the average or overall brightness of the image, and the window *width* controls the gray scale, (the inverse of contrast). This means that windowing the image is simply a *re-application of gradation processing,* or later alterations made by the operator that use the same computer algorithms as default gradient processing does. The specific applications of windowing were discussed in Chapter 4, but the following discussion explains the "how" behind these applications.

During gradation processing, the "Q values" of the rescaled data set are fed into an anatomical look-up table or LUT. An *anatomical LUT* is stored by the computer for each specific type of radiographic procedure. When the radiographer enters a particular procedure at the console, the anatomical LUT is automatically selected. It will *customize* the gray scale and brightness of the image according to the specific anatomy and most common pathologies expected to be displayed.

As shown in Table 6-1 in the previous chapter, LUT's really are *tables,* not graphs, but a graph such as the one in Figure 7-5 helps us *visualize* what the LUT is doing mathematically. It also represents the actual *formula* or algorithm used to do it in the form of the *function curve* depicted (*Fx* in the graph). The rescaled Q values for the image are shown at the bottom of this graph, and are being entered into a formula that the function curve represents. When we extend a vertical line upward from any one of these Q values, it is *reflected* off the function curve horizontally to the right, where we can read the output or final pixel value that resulted.

The formula represented by the function curve in Figure 7-5 results in enhanced contrast. We can measure the contrast of the original input image by finding the ratio (dividing) between two selected input values, such as 13 divided by 10 to obtain an original image contrast of 1.3. We see what the computer algorithm does with these numbers by extending a vertical line from each of them up to the function curve, and then

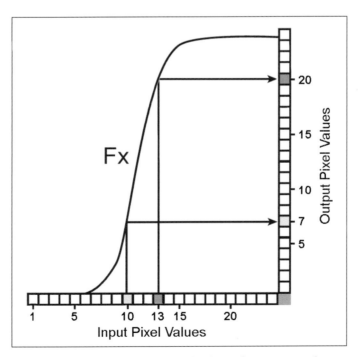

Figure 7-5. The gradation processing curve *Fx* represents the formula acting on the input pixels. Here, this function increases image contrast from $13/10 = 1.3$ to $20/7 = 2.9$. (From Q. B. Carroll, *Radiography in the Digital Age,* 3rd ed. Springfield, IL: Charles C Thomas, Publisher, Ltd., 2018. Reprinted by permission.)

to the right over to the output scale. Here, we find that the algorithm formula has reduced the pixel value 10 down to 7, but the pixel value 13 has been increased to 20. The contrast of the new output image is 20/7 = 2.9. Contrast has been enhanced substantially, from 1.3 to 2.9.

Most gradation processing operations use formulas called intensity transformations. Here are three examples:

Gamma Transformation: s = cr^{γ}
Log Transformation: s = $clog(1 + r)$
Image Negative: s = L – 1 – r

In each of these formulas, *r* is the input value for any pixel and *s* is the output value resulting from the formula. In the *gamma transformation* formula, *c* is a constant usually set to 1.0 and thus can be ignored for our purposes here. We then see that the output value *s* is calculated simply by raising *r* by the exponent *gamma* (γ). Figure 7-6 demonstrates that as *gamma* is reduced to fractions of 1.0 for this MRI image of a lateral thoracic spine, the gray scale is lengthened, thus contrast is reduced. Figure 7-7 shows aerial photographs of an airport which have been subjected to increasing values of *gamma* above 1.0, and one can readily see that the contrast is increasing.

The intensity transformation formulas above are just three examples of many dozens of formulas used by manufacturers for gradation processing. Each can be represented by a function curve graph. Figure 7-8 illustrates just four of FujiMed's 26 types of function curves for gradation processing.

For gradation processing, many different mathematical approaches can be used to accomplish a particular task such as increasing image con-

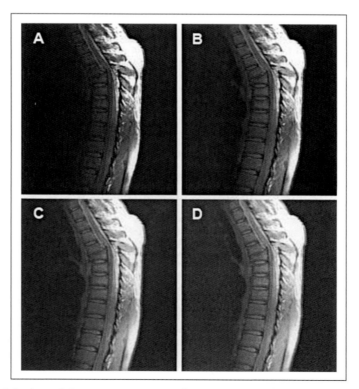

Figure 7-6. MR images of the lateral thoracic spine using the gamma transformation formula. For *A*, gamma = 1.0, *B* = 0.6, *C* = 0.4, and *D* = 0.3. Decreasing gamma lengthens the gray scale. (From Q. B. Carroll, *Radiography in the Digital Age,* 3rd ed. Springfield, IL: Charles C Thomas, Publisher, Ltd., 2018. Reprinted by permission.)

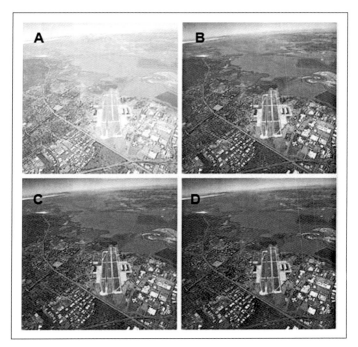

Figure 7-7. Aerial photographs of an airport using the gamma transformation formula. For *A*, gamma = 1, *B* = 3, *C* =4, and *D* = 5. Increasing gamma shortens the gray scale, increasing contrast. (From Q. B. Carroll, *Radiography in the Digital Age,* 3rd ed. Springfield, IL: Charles C Thomas, Publisher, Ltd., 2018. Reprinted by permission.)

trast. X-ray machine manufacturers keep the specific algorithms they use as patented trade secrets. We would expect some of these to be more effective than others, and this is the case. (This can be a source of frustration for the student, because it means that when a particular experiment is done with digital lab equipment, one cannot assume that other manufacturers equipment will produce identical results, especially in terms of effectiveness.)

Figure 7-9 is a graphical example of how changing one of the parameters in a formula (like *gamma*) alters the function curve and the resulting image. This is an actual example using Fuji's parameter called *GA* (for *gradient amount* or *gradient angle*). We see that when *GA* is greater than 1, the resulting function curve is steeper than 45 degrees, and image contrast is increased, (compare to Figure 7-5), but when *GA* is less than 1, the resulting function curve is shallower than 45 degrees and image contrast is reduced as gray

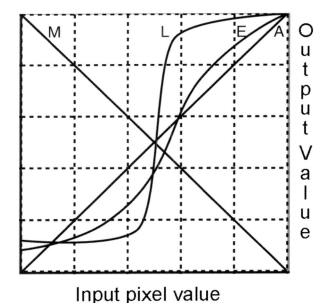

Input pixel value

Figure 7-8. Four examples of FujiMed's 26 function curves for gradation processing. Curve *M* reverses the image to a positive "black bone" image. (Courtesy, *FujiMed, Inc.*)

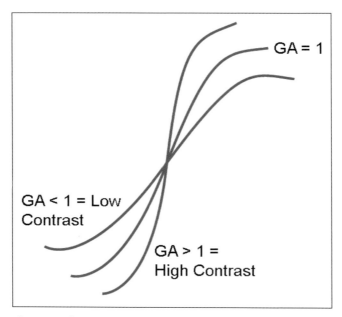

Figure 7-9. Example of how changing the intensity transformation formula affects the graphed function curve. Here, FujiMed's *GA function* (gradient angle) rotates to a low-contrast slope when GA is set to less than 1, or to a high-contrast slope when GA is set to greater than 1. (Courtesy, *FujiMed, Inc.*)

scale is lengthened. Fuji uses 26 different types of formulas in their gradation processing software, resulting in 26 different shapes of function curves.

Thousands of calculations using the original formula do not need to be done on each image. Rather, the formula (or function curve) is used to first construct an actual look-up table, an anatomical LUT for a particular type of radiographic procedure, such as the one in Table 6-2 in the previous chapter. This anatomical LUT is then permanently stored. When a specific *image* is processed, input pixel values are simply fed into the anatomical LUT and the corresponding output values are read out from the table, a process that can be executed in a miniscule fraction of a second.

Default gradation processing, using anatomical LUTs, is performed on every image before it is initially displayed on a monitor. Windowing level and window width adjustments made later by the operator, (*levelling* and *windowing*) are effectively adjustments *to the anatomical LUT,* that is, they are a re-application of gradation processing.

Whether it is part of default processing or part of later windowing adjustments, gradation processing is always considered as postprocessing because it involves refinements to the image based on anatomy and specific diagnostic needs, rather than corrections to flaws in image acquisition.)

Data Clipping

In Chapter 1 we defined *bit depth, dynamic range,* and *gray scale.* There must be "room" within the dynamic range to both display an image with its full gray scale *and* allow for windowing adjustments upward and downward in both brightness and contrast. *Data clipping* can occur when either the bit depth of the hardware or the dynamic range of the system are too limited. Figure 7-10 shows graphically how this might occur for a system that is 8 bits deep for a dynamic range of 256 "shades of gray," and the window level is increased to darken the image. The gray scale of the adjusted image is represented by the dashed line. Before it reaches its original length (equal to that of the solid line), it

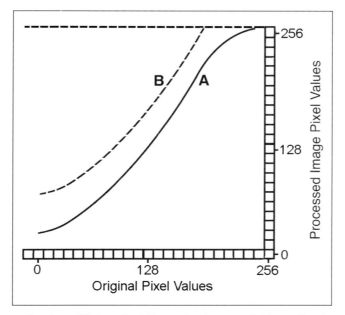

Figure 7-10. Data clipping via leveling: With only 256 pixel values available in the dynamic range, increasing the window level to darken the image raises the gray scale curve to *B* where it is truncated at the darkest level of 256 (*at the top of the graph*). Data is lost.

"runs out" of available pixel values at 256. Values darker than this cannot be produced by the system. The line is truncated, representing the loss of the darkest pixel values from the image.

Figure 7-11 shows a similar result when the contrast is increased for this system limited to 256 values. Note that once again, the dashed line is shorter than the original solid line. As the gray scale was increased, it was unable to extend pixel values beyond 256, and pixel values (data) are lost from the image. Data clipping limits the radiologist's ability to window the image, and can adversely affect diagnosis. This is one good reason why radiographers should generally not *save* images that they have windowed into the PACS, replacing the original image.

Dynamic Range Compression

Dynamic range compression (DRC) is the removal of the darkest and the lightest extremes of the pixel values from the gray scale of a digital image. Since the bit depth of a typical computer far exceeds the range of human vision, some computer storage space can be saved by not using the entire bit depth of the system, Figure 7-12*A*. When DRC is used for this purpose, enough dynamic range should be left to still allow the actual displayed image to be windowed up and down as shown in Figure 7-12*B* to darken or lighten it. That portion of the dynamic range actually being displayed is the gray scale of the image itself.

Most radiographic techniques are designed to primarily demonstrate bony anatomy or contrast agents. For conventional film radiography, this often resulted in soft tissue areas of the image being displayed quite dark. In some cases, the radiographic procedure was ordered to rule out very subtle foreign bodies such as a chicken bone lodged within an esophagus, small glass slivers, or wood slivers. To demonstrate these types of objects, *soft tissue techniques* were developed which typically consisted of reducing the kVp by up to 20 percent from the usual technique for that anatomy, with no compensation

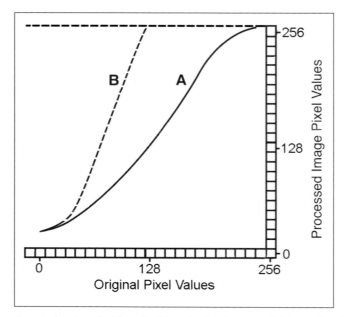

Figure 7-11. Data clipping via windowing: With only 256 pixel values available in the dynamic range, decreasing window width to increase contrast raises the *slope* of the gray scale curve to *B* such that it is again truncated at the darkest level, 256 (*at the top*). Again, Data is lost.

in mAs. This both lightened the image from reduced exposure and brought the penetration of the x-ray beam down to a level more consistent with the meager x-ray absorption of soft tissues. Fat pads demarcating the edges of muscles or ligaments important to the orthopedic surgeon could also be demonstrated. However, bones in these images were then depicted too light.

Remember that *rescaling* is designed to always produce an image with the same type of overall

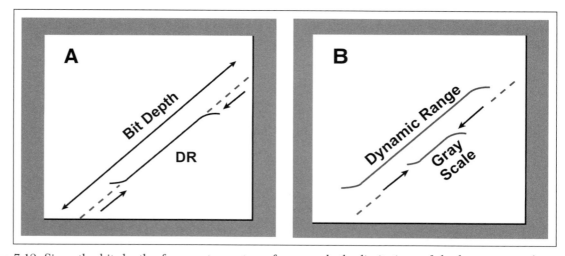

Figure 7-12. Since the bit depth of computer systems far exceeds the limitations of the human eye, *A*, computer storage space can be saved by setting the dynamic range much less than the bit depth without sacrificing image quality. Enough dynamic range must be left intact to allow the actual gray scale of the image to be windowed up and down as needed for all diagnostic purposes, *B*. (From Q. B. Carroll, *Radiography in the Digital Age*, 3rd ed. Springfield, IL: Charles C Thomas, Publisher, Ltd., 2018. Reprinted by permission.)

density and contrast, typically geared toward the demonstration of bony anatomy. If the conventional soft tissue technique were used, digital processing would simply compensate for it, cancelling out the desired effect. Therefore, in the digital age, soft tissue technique must be simulated through special features in digital processing rather than with the radiographic technique used for the original exposure. Dynamic range compression can do this.

The dynamic range can be compressed to such a degree that it begins to affect the gray scale of the displayed image. Figure 7-13 shows this as a truncation of the gray scale that "clips off" the toe and the shoulder portions of the gray scale curve, or the darkest and the lightest extremes of the pixel values in the *displayed* image itself. The darkest darks, and the lightest light shades of the gray scale are *no longer available* for constructing the image. From the midpoint of the gray scale, the computer has progressively reduced the pixel values above, and progressively increased the pixel values below, this point. Extremely dark areas are lightened, and extremely light areas are darkened. Thus,

several manufacturers use the term *tissue equalization* or *contrast equalization* for this processing feature. Because the darkest portions of the image are lightened up, subtle slivers or chicken bones are better visualized, mimicking the old soft tis sue techniques, but with the advantage of still maintaining proper density for the bones. DRC can thus aid in diagnosis. Shown in Figure 7-14, the resulting image has an overall *grayer* appearance after DRC has been applied, yet this is not a bad thing, since more tissues of all types are being demonstrated, e.g., more details are visible within the lighter heart shadow and also within the darker lung tissues.

The concept of dynamic range compression can be confusing for the radiography student because it seems to violate the axiom that "shorter gray scale increases contrast." The key is to visualize conventional shortening of the gray scale as "squeezing" a set range of densities that *still extends from white to black,* whereas DRC actually *cuts off* the ends of the range of densities as shown in Figure 7-13. The resulting image appears grayer because these extremes in density are *missing.* The term "compression" is quite mis-

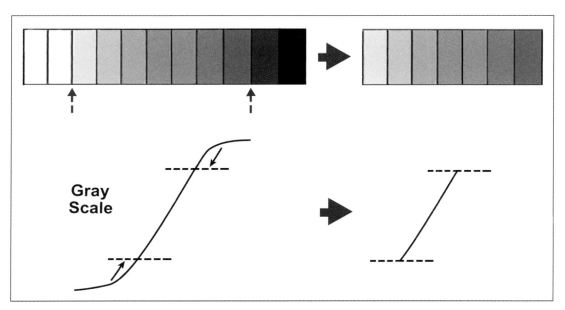

Figure 7-13. When applied to the degree that it visibly affects the final image, dynamic range compression (DRC) truncates the gray scale, "cutting out" the darkest and the lightest densities in the displayed image (see Figure 7-14).

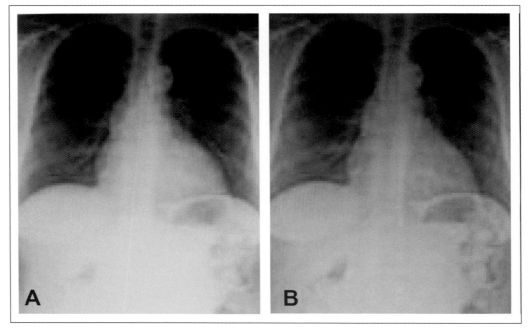

Figure 7-14. Compared to image *A*, Fuji's *dynamic range control (DRC)* applied to chest image *B* shows structures through the heart shadow now visible due to darkening, while in the lower lung fields more vascular and bronchial structures are now visible due to lightening. This is an example of *tissue equalization* or *contrast equalization*. (Courtesy, *FujiMed, Inc.*)

leading here, and this process would better have been named *dynamic range truncation.*

Fuji's *dynamic range control* is dynamic range compression combined with other algorithms that customize the gray scale according to the anatomy to be demonstrated for each type of radiographic procedure.

Chapter Review Questions

1. Digital image processing always begins and ends in the _____ domain.
2. Pixels are operated upon based on the pixel values in the _____ domain.
3. In the frequency domain, instead of identifying pixels, _____ are identified within the image according to their size.
4. Small objects are associated with _____ frequencies.
5. Sorting an image by spatial location results in a(n) _____.
6. Sorting an image by the intensity of the pixel values results in a(n) _____.
7. A submatrix that is passed over the larger image matrix executing some mathematical function is the definition of a(n) _____.
8. All kernel operations take place in the _____ domain.
9. List the three categories of spatial domain operations:
10. All image reorientations are _____ processing operations in the spatial domain.
11. Histogram analysis and all gradation processing operations are in the _____ domain.
12. A gradient curve with a steep slope represents an image with _____ contrast.
13. Leveling and windowing done by the operator are actually reapplications of _____ processing.
14. When the radiographer enters a particular anatomical procedure at the console,

an anatomical _____ is automatically selected for default gradation processing.

15. The *function* curve for gradation processing actually represents a _____ to which the pixel values are being subjected.

16. Most gradation processing operations use mathematical formulas called _____ transformations.

17. When the dynamic range of an imaging system is too limited, or when a radiographer subjects the image to extreme windowing and then *saves* the windowed image, data _____ can occur.

18. When *dynamic range compression* is applied to such an extent that it affects the visible displayed image, it results in _____ equalization in which the darkest densities are lightened and the lightest densities are darkened.

19. Although *dynamic range compression* results in a "grayer" looking overall image, this effect is *not* due to lengthening the gray scale, but actually to _____ _____ the darkest and lightest densities in the gray scale.

20. The diagnostic advantage of *dynamic range compression* is that _____ details become visible in areas of the image that were previously too dark or too light.

Chapter 8

DEFAULT POSTPROCESSING II: DETAIL PROCESSING

Objectives

Upon completion of this chapter, you should be able to:

1. Define *detail processing* and list the two domains in which it can operate.
2. Describe how the image on a display monitor can be represented as frequencies.
3. Conceptualize how frequencies can be separated or recompiled with *Fourier transforms*.
4. Define *multiscale processing, band-pass filtering* and *unsharp-mask filtering*.
5. Explain how *smoothing, edge enhancement,* and *background suppression* are achieved both by frequency processing and by using kernels.
6. Describe two ways in which the final digital image must be formatted before display.

Detail processing is a concept unique to digital technology, in which structures in the image can be selected according to their *size* and singled out for contrast enhancement or suppression. For example, very small details can have their *local contrast* increased, making them stand out more against the background of mid-size structures and larger tissue areas. Alternatively, mid-size structures can be selected to be *suppressed* more into the "background" by contrast reduction. For technologists who had worked with film, this would seem miraculous—film emulsions and chemical processing could only adjust the density and contrast of the overall image as a whole.

The term *detail processing* stems from the ability of digital algorithms to treat the fine details of an image *separately* from the larger gross structures in the image. Recall that in the frequency domain, the image is first sorted out according to the *size of the objects* in it rather than by anything to do with pixels. With the smallest details throughout the image sorted into a separate computer file, their contrast (for example) can be enhanced while leaving the general contrast of the overall image untouched. Upon close inspection of the resulting image, these fine details will be more visible against the background, yet when one stands back at a distance and examines the image as a whole, the overall contrast appears about the same as it was before detail processing was applied. This is referred to as *decoupling local contrast from global contrast.*

The results seem almost mystical, yet they can be accomplished in *two* of the three processing domains—the frequency domain and the spatial domain. We will first examine the approach that is least intuitive to understand, frequency processing, then come back to see how the same types of effects can also be achieved in the spatial domain using special tools called *kernels.*

Understanding the Frequency Domain

It is useful to review from Chapter 3 the section on *Resolution at the Microscopic Level.* When you examine the *density trace diagram* in Figure 3-11, and use a little imagination, it resembles a wave moving up, down, then up again. Look at the illustration of *modulation transfer function (MTF)* in Figure 3-12 and again, we see the appearance of waveforms that represent alternating densities within the latent image.

We can apply similar concepts to the digital image displayed on a viewing monitor. First, imagine a single row of pixels on the monitor, alternating between black and white. As shown in Figure 8-1, this pattern can be represented as a *waveform* alternating up and down. We can choose to have the top peaks of the wave represent either the brightest (white) pixels or the darkest (black) pixels. Here, the peaks or crests indicate dark pixels and the troughs or dips in the wave represent light pixels. Note that for pixel row *B*, the resulting waves are not as high because there is less difference between gray and white than there is between black and white. We begin to see that the *amplitude* of these waves, how "tall" they are, represents the difference in *brightness* or *density* between pixels. (This is the contrast between pixels.)

Now let's examine the *frequency* of the alternating signal from the pixels. The zero-point of each individual wave corresponds to the transition border between each pair of pixels. Therefore, the *wavelength* of each pulse represents the width of the pixel, that is, the pixel size. When the brightness of the darker pixels was changed between *A* and *B* from black to gray, the pixel size did not change, nor did the wavelength. We can think of the *frequency* of this waveform as the number of up-down cycles across an *entire row* of pixels, in other words, from the left edge of the display monitor screen to the right edge. If there are 600 pixels in a particular row, the frequency is 300 cycles or 300 *hertz (Hz)*. Each up-down cycle consists of two *pulses* (one up, one down), and each pulse represents one pixel. So, the measured frequency in cycles is always one-half the number of pixels in a row. In Figure 8-1, since the wavelength is unchanged between *A* and *B*, the frequency is also unchanged.

If the pixels themselves could be made smaller, more would fit across the same row, resulting in a higher frequency. The frequency of a row of hardware pixels depends on the *size* of the pixels.

Objects within the image also have a frequency related to the number of pixels they *occupy* in each row: If an object and a space of identical size can just fit across the display screen from left-to-right, they have a lateral frequency of 2 hertz, if 5 such objects and spaces can fit across the screen, their frequency is 5 hertz. The 5-

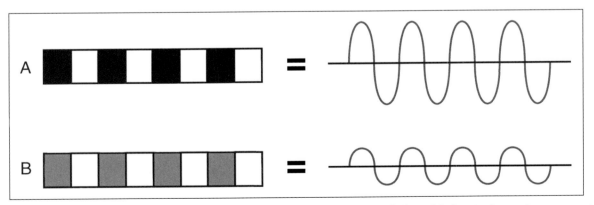

Figure 8-1. Along a single row of pixels on a display monitor, alternating dark and light pixels can be represented as *up* and *down* pulses in a waveform. Here, from *A* to *B*, the wavelength is the same because pixel size does not change, but the lighter gray pixels in *B* create waves with lower amplitude.

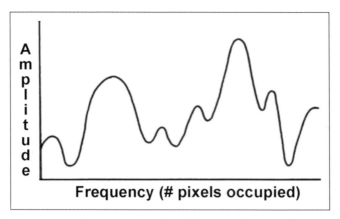

Figure 8-2. Complex waveform representing one row of pixels in a real image. Longer wavelengths represent larger objects (occupying more pixels), taller waves represent darker pixel values. Wavelength correlates to frequency, wave height is amplitude. (From Q. B. Carroll, *Radiography in the Digital Age,* 3rd ed. Springfield, IL: Charles C Thomas, Publisher, Ltd., 2018. Reprinted by permission.)

hertz objects must be smaller than the 2-hertz objects. *Large objects make large wavelengths, and they are low-frequency objects because fewer can "fit" across the screen. Small objects make short waves, and they are high-frequency objects because a high number of them will "fit" across the screen.*

Of course, in a real clinical radiographic image, across a selected row of pixels there will be objects or structures of *various sizes,* so the actual waveform for a row of pixels is complex and looks more like Figure 8-2. This graph plots the amplitude (height) of each wave against its frequency (or wavelength). Amplitude (height) represents the gray level or pixel value being displayed, whereas wavelength represents the *number of pixels* the object occupies across the row, hence, its lateral size. Taller waves represent darker objects, *wider waves represent larger objects* which occupy more pixels across the row. For example, a tall, skinny spike would represent a dark but small object.

The ability of frequency processing to *separate* structures according to their size is provided by a mathematical process called *Fourier transformation* which can effectively break this complex waveform into its component waves, long, medium, and short. Figure 8-3 illustrates how a complex wave can be represented as the *sum* of pulses having different wavelengths, each representing a different object. The complex wave is

formed when these wave pulses are *superimposed* over one another, effectively adding all the different objects together that make up a particular row in the image.

This process follows exactly the same rules that one would observe with water waves: If two water waves merge together with "perfect timing," such that their crests align, adding water upon water, the result will be a single wave that is taller. However, if the crest of one wave aligns with the trough of the second wave as they merge, there will be a "cancelling out" effect in which the resulting wave is much diminished or levelled. Shown in Figure 8-4, *A,* if a large positive (up) pulse is superimposed over a smaller positive pulse, the result looks like the large wave with an added "bump" at the top. In Figure 8-4, *B,* we see the effect of adding a small *negative pulse* or *trough* to the large positive pulse, a "dip" in the larger wave. This is just how a complex waveform is created, and gives us a hint as to how it can also be "taken apart" for analysis by Fourier transformation.

Now examine Figure 8-3 one more time with frequencies in mind: The original image matrix is 10 pixels across. For simplicity, we have broken the waveform for this row of pixels into just 3 frequencies, 1.25 hertz, 2.5 Hz, and 5 Hz. Let us define an "object" or "detail" as a single pulse. For the small 5 Hz details in row D, 10

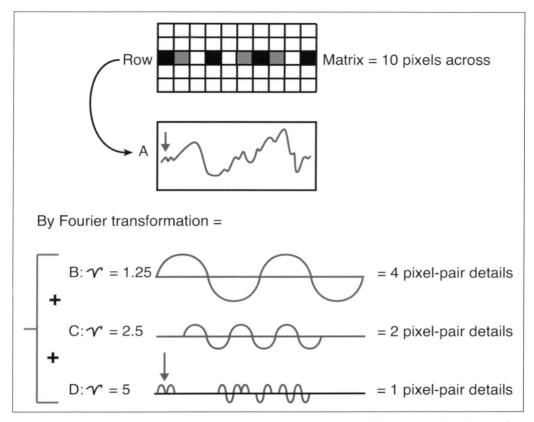

Figure 8-3. A complex wave can be represented as the *sum* of pulses with different wavelengths, each wavelength representing different *sizes* of objects or structures in one row of the image. Here, frequencies (wavelengths) **B**, **C** and **D** add up to form the complex wave **A**. A complex wave can be represented as the *sum* of pulses with different wavelengths, each wavelength representing different *sizes* of objects or structures in one row of the image. Here, frequencies (wavelengths) **B**, **C** and D add up to form the complex wave A. (From Q. **B**. Carroll, *Understanding Digital Radiograph Processing* (video), Denton, TX: Digital Imaging Consultants, 2014. Reprinted by permission.)

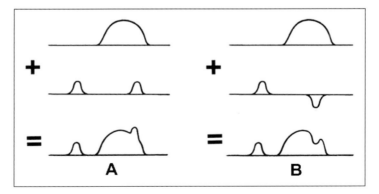

Figure 8-4. For waves, if a small positive pulse is added to a larger positive pulse, **A**, the result is a large pulse with an added "bump." If a small *negative* pulse is added to the larger wave, a "dip" in the large pulse appears, **B**. This is how complex waves form. (From Q. B. Carroll, *Understanding Digital Radiograph Processing* (video), Denton, TX: Digital Imaging Consultants, 2014. Reprinted by permission.)

pulses or 10 of these "objects" could be recorded across the display screen. However, you will note that there are breaks between these graphed pulses. This is because these small objects only occur in certain places in the image, not at every pixel. The positive (up) pulses represent dark details, the negative (down) pulses represent light details of the same size.

Line *C* graphs objects with a frequency of 2.5 Hz or 2.5 up-down cycles that would fit across the screen. But, each cycle has two pulses, and we've defined an "object" as one pulse, so we must multiply by 2 to find that 5 of these objects would be able to fit across the display screen. There are 3 dark objects and 3 light objects of this size present, but note that the waveform does not begin at the far left, that is, there is no object of this size at the far left of the image. Line B graphs objects of 1.25 Hz. Doubling this amount, we see that 2 of these objects would fit across the screen, and they fill the screen from left-to-right.

When lines *D*, *C* and *B* are superimposed (summed), we obtain the complex waveform in line *A* which represents the actual image across this one row of pixels. The arrow at the far left shows where the two small objects in line *D* are added to the large pulse (object) in line *B*. Immediately to the right of these two little pulses in *A*, we see a broader, large peak. This is the sum of the first "mid-size" pulse in line *C* and the *right half* of the first large pulse in line *B*.

Frequency Detail Processing

Figure 8-5 illustrates the actual image effects of Fourier transformation breaking the original image into its component frequencies. We think of these as the "frequency layers" of the image. Layers B through D are high-frequency layers and demonstrate only the finest details, especially in the bone marrow. Note that the dark background density is missing. This is because it occupies a large area of the image and is therefore treated as a large "structure" with very low frequency that does not belong in this layer. Layers E and F are mid-frequency layers and demonstrate such structures and the cortical

portions of the bones. Layers G through I are low-frequency layers demonstrating the largest structures in the image, the soft tissue of the hand and the background density.

The term *multiscale processing* refers to decomposing the original image into eight or more frequency layers, performing various operations on selected individual layers, and recomposing the image. Multiscale processing software was pioneered by Agfa as *MUSICA (Multi-scale Image Contrast Amplification)*, by Philips Healthcare as *UNIQUE (Unified Image Quality Enhancement)*, and by FujiMed as *MFP (Multi-objective Frequency Processing)*. The original image is repeatedly split into a high-frequency component and a low-frequency component. Each time, the high-frequency image is set aside, and the low-frequency component is subjected to the next division, until there are eight or more image layers.

While these layers are separated, any one of them can be operated upon by various computer programs to enhance or suppress the visibility of those structures in the image, without affecting structures of other sizes. Figure 8-6 illustrates how enhancement consists of simply boosting the signal for everything in that computer file (multiplying all pixel values in that layer by some constant factor). A single layer can also be subjected to gradation processing and any number of other treatments. After all of these operations are completed, the image layers are compiled back together to form the final displayed image. This is done mathematically using *inverse Fourier transformation*, which adds all the various waves back together (Figure 8-7). (Remember that these are actually mathematical processes, not graphical ones, but graphs are used in these figures to help visualize what is happening.)

The reconstituted complex waveform in Figure 8-7 represents one single row of pixels in the reassembled image for display. The peaks and valleys of the waveform will determine the amount of *amperage* or *voltage* applied to each pixel in the row, resulting in the different levels of emitted brightness that will constitute the final displayed image.

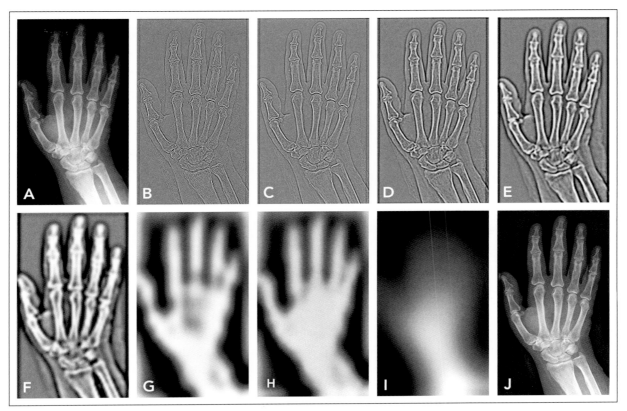

Figure 8-5. Decomposition of an image of the hand, *A*, by Fourier transformation into eight component frequency "layers" from highest (*B*) to lowest (*I*). At the highest frequencies, *B* through *D*, only the finest details such as bone marrow trabeculae are seen. At mid-level frequencies *E* and *F*, cortical bone structures become apparent. Large masses of soft tissue are brought out in *G* and *H*. Only at the lowest frequency, *I*, are the large areas of surrounding "background" density from raw radiation exposure to the IR are demonstrated. *J* is the reconstructed image. (Courtesy, Dr. Ralph Koenker, Philips Healthcare, Bothell, WA. Reprinted by permission.)

A selected layer can also be entirely left out upon reconstructing the image, Figure 8-8. This process is called *band-pass filtering*. It is widely used in noise reduction, especially for electronic mottle which has a consistent size. For example, suppose there are several white spots in an image due to dexel drop-out from an aged image receptor that was used to acquire the image. We desire to filter out this noise in the image.

A *low-pass filtering* algorithm will keep, or "pass through," the low-frequency layers in the image. By implication, one or more high-frequency layers will be removed by leaving it out upon reconstruction. The electronic mottle we have just described will consist of very small white spots whose size can be reasonably pre-

dicted, since they will be the size of 2 or 3 adjacent pixels, one of which was supposed to be black or gray and is not. Therefore, the frequency layer that corresponds to this size of object, say, 0.2 to 0.3 mm, is the layer we will target for removal. On the console controls of an x-ray machine, this feature will be most commonly called a *smoothing* function, but may have various proprietary names for different manufacturers. An example of smoothing is presented in Figure 9-6 in Chapter 9.

As the student might surmise, there is a price to be paid for deleting a detail layer from the image. Some of the finer diagnostically useful details in the image, such as small bony structures of the same size, 0.2 to 0.3 mm, will also

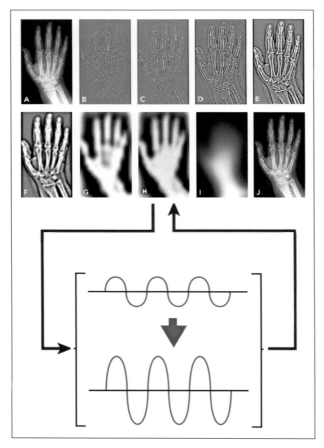

Figure 8-6. During frequency processing, any frequency "layer" can be selected, subjected to enhancement, suppression, or other operations, and then reinserted into the image "stack." Here, the amplitude of layer *H* is boosted, enhancing contrast *only* for objects sized to that layer.

be lost. The elimination of noise from the image generally requires that we accept some loss of fine detail along with it.

High-pass filtering, so named because it "passes through" the highest frequency layers, targets one or more mid- or low-frequency layers to be removed. *Background suppression* refers to elimination of the very lowest frequencies. In both of these cases, small details and the fine edges of structures visually will stand out better against the surrounding larger anatomy. For the radiographer, perhaps the best generic clinical name for high-pass filtering is *edge enhancement,* yet again, different manufacturers will use different proprietary names for it, sometimes justified

because it may be combined with other processes to achieve a more general desired improvement to the image.

For example, General Electric has a feature called the *look* of the image. Both the console setting and the resulting images are shown in Figures 9-7 and 9-8 in Chapter 9. The operator can select a "soft" look, the "normal" look of the original displayed image, or a "hard" look. In this program, the "soft" look is achieved essentially by smoothing algorithms, and the "hard" look is achieved primarily by edge-enhancement algorithms, although other algorithms may also be part of the program.

CareStream Health has a default detail processing suite that includes three steps: 1) edge enhancement, 2) *enhanced visualization processing* or *EVP,* and 3) *perceptual tone scaling (PTS).* For the EVP stage, the image is split into just two layers, the contrast of the low-frequency layer is suppressed, and the contrast of the high-frequency layer is increased. The PTS stage is really a variation of *gradation processing* but uniquely modeled on "psycho physical" studies of human visual perception. The variations of processing features among all the manufacturers of x-ray equipment can become overwhelming, but the essential processes are the same and are described here with their most common or most generic names.

Transition Between Image Domains

In the previous two chapters, we learned that the image must be transitioned from the *spatial domain* into the *intensity domain* so that histogram analysis and gradation processing can be performed. During this time, the computer must keep a record of *where* the pixels must be reinserted into the spatial matrix to reconstruct the image after all these intensity operations are completed.

In a similar way, *frequency detail processing* requires that the image be transitioned from the spatial domain into the *frequency domain,* where all the detail processing operations are executed, then placed back into the spatial matrix after detail processing is completed. In this case, the

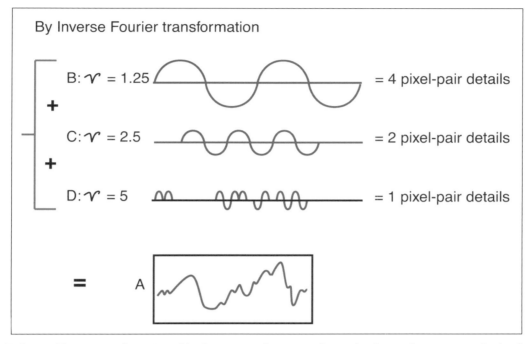

Figure 8-7. *Inverse Fourier transformation* adds the various frequency layers back together to compile the final image for display. The reconstituted complex waveform **A** controls the amperage or voltage applied to each pixel in a particular row of the image matrix, resulting in different levels of displayed brightness. (From Q. B. Carroll, *Radiography in the Digital Age,* 3rd ed. Springfield, IL: Charles C Thomas, Publisher, Ltd., 2018. Reprinted by permission.)

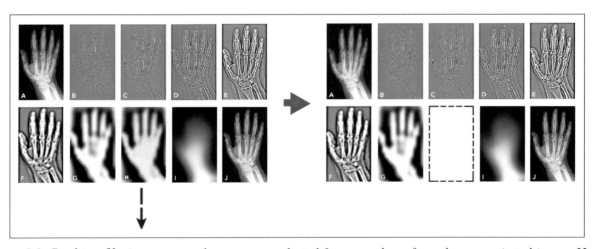

Figure 8-8. *Band-pass filtering* permanently removes a selected frequency layer from the reconstituted image. Here, a low-frequency layer (**H**) is left out upon reconstruction. This is *high-pass* filtering, since all the high-frequency layers were "passed through" to the final displayed image intact. Removing low-frequency layers results in edge enhancement, removing high-frequency layers results in smoothing or noise reduction.

location of each object in the image must be separately stored as a set of designated pixel locations while all the detail operations are going on. After all the frequency operations are completed, the computer will have a log of where to place each object in reconstructing the image in its original spatial matrix.

The following section describes *detail processing by kernels.* Since a kernel is a small *matrix,* it is a spatial operation. No transition between processing domains is needed, so it is not necessary to create a log of pixel or object locations. Kernels are directly applied to the spatial image in its original matrix.

Spatial Detail Processing (Kernels)

A *kernel* is a "submatrix", a small matrix that is passed over the larger matrix of the whole image in order to change all of the pixel values mathematically. In Chapter 5 we discussed how a kernel can be used to correct for dexel drop-out defects in the incoming image as part of pre-processing. A kernel can also be used for detail processing when simple *multiplication factors* are placed in its cells.

In Figure 8-9, we see three examples all using small kernels that consist of only 9 cells, or a 3 x 3 cell matrix. In practice, most kernels are larger in size but must always have odd-numbered dimensions, e.g., 5 x 5, 9 x 9, and so on such that a *center cell* can always be identified. This is necessary so that as the kernel is passed over the image, the kernel's *location* can always be identified by the image matrix pixel that the kernel is centered over. (With even-numbered dimensions, the kernel has no cell that can be identified as the center cell of the kernel.)

Beginning at the top-left pixel in the image, the kernel is passed from left-to-right along the entire first row of pixels, then indexes down to the next row and repeats another left-to-right sweep until every row has been covered (see Figure 7-2 in Chapter 7). This process is then repeated *vertically,* with the kernel sweeping downward along the first *column* of image pixels, indexing right to the next column, and sweeping downward again until all columns have been covered.

During these movements, the kernel pauses over each individual pixel in the image, and, while centered over that pixel, multiplies all of the pixels in the neighborhood by the values inserted in the kernel cells. To understand what occurs at *each pixel in the image,* see Figure 8-9 and examine the middle row in kernel *B*: As this kernel moves left-to-right, a particular pixel in the image will first have its value multiplied by −1 in the right-most kernel cell. When the kernel moves to the right again, this same pixel value will be multiplied by 9, then after the next movement, it will again be multiplied by −1 from the left-most kernel cell. When this kernel indexes down to the next *row* of pixels and pass-

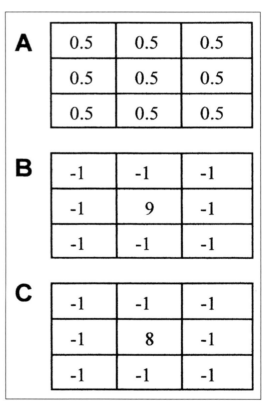

Figure 8-9. Some different types of *kernels: A,* with all values in the cells equal, results in smoothing and noise reduction. The cell values in *B* sum to 1, resulting in edge enhancement. The values in *C* sum to zero, resulting in background suppression. (From Q. B. Carroll, *Radiography in the Digital Age,* 3rd ed. Springfield, IL: Charles C Thomas, Publisher, Ltd., 2018. Reprinted by permission.)

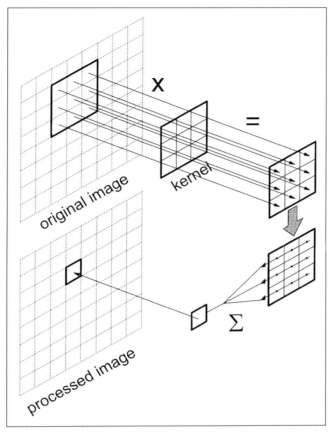

Figure 8-10. Detail processing with a kernel. For each step, the image pixel values covered by the nine cells in the kernel are multiplied by the kernel values (top), then these nine products are summed to insert the final value into the centered pixel (bottom). The kernel then indexes to the next pixel and this operation is repeated for all rows and all columns. (From Bushberg et al., *The Essential Physics of Medical Imaging*, 2nd ed., Baltimore: Williams & Wilkins.)

es left-to-right, this same pixel will be subjected to the top row of factors in the kernel. In this row, the pixel value will be multiplied by –1 three times.

During horizontal sweeps, a particular image pixel is treated by all 9 cells in the kernel. All nine of these operations will be repeated when the kernel passes over the image again vertically. So, for kernel **B**, each pixel in the image will undergo a total of 18 multiplications. But, this is not all. *Each time the multiplications take place, the resulting products from all 9 pixels under the kernel are then added together, and it is this sum of the products that is finally inserted into the centered pixel,* shown in Figure 8-10. Each step consists of multiplying all 9 pixel values under the kernel by the respective factors, then summing these 9 products.

The resulting effect on the radiographic image depends on the *relationship* between the multiplying factors in the kernel. In Figure 8-9, kernel **A** has equal core values throughout. As long as these numbers are greater than zero, the result of this type of kernel will be *smoothing* of the image; Noise will be reduced, but with an attendant slight loss of contrast at detail edges. Kernel **B** contains both positive and negative numbers, but note that if they are all added together, they sum to 1.0. This type of kernel results in *edge enhancement,* increasing local contrast at the edges of structures and for small details. Kernel **C** also presents both positive and negative cell values, but these sum up to zero. This type of kernel results in *background suppression,* a reduction of contrast only for large structures in the image, leaving the foreground details more visible.

Kernels can also be applied to the global (overall) contrast of an image, which is increased when both positive and negative core values sum to a number greater than 1, and decreased when they sum to a number less than 1. However, gradation processing is more widely used for this purpose.

For every detail processing effect that can be achieved in the frequency domain, there is an equivalent process in the spatial domain that can achieve the same type of result. By adjusting the size of the kernels used, we can achieve the effects of *band-pass filtering,* effectively removing certain frequency ranges (layers) from the image. In other words, we can target a threshold *size* of detail or structure to be eliminated throughout the image. The larger the kernel (the more cells in it), the larger the structures that will be suppressed. The kernel size must be carefully selected according to the type of anatomy being imaged. Too large a kernel will be ineffective in achieving the desired enhancement, while too small a kernel may remove important details from the image such that diagnostic information is lost.

Separation of gross (large) anatomy into a separate layer can be accomplished by using a kernel that effectively averages local pixel values. If this gross image is then *reversed* to create a positive (black-on-white) copy, it can then be *subtracted* from the original by superimposing the two. Gross structures will be cancelled out between the negative original and positive "mask" images, black cancelling white, white cancelling black, dark gray cancelling light gray, and light gray cancelling dark gray, such that these gross structures nearly disappear. Meanwhile, the mid-size structures and fine details in the image are not cancelled, because they are not present in the reversed positive "mask" image. In effect, the result is an image that has undergone *edge enhancement* (as with *high-pass frequency filtering*), but using spatial methods.

This type of processing has been misleadingly named "unsharp mask filtering" because the mask image, having only gross structures in it, has a blurry or unsharp appearance to it. However, it is not truly blurred in any geometrical sense. It only has this subjective appearance because there are no small details or fine edges present in it. A more accurate, if cumbersome, name for this process would be "gross-structure mask filtering."

As shown in Figure 8-9*A*, kernels can also be configured to achieve a *smoothing* effect, (the opposite of edge enhancement), which has the added bonus of reducing noise. As described in Chapter 5, kernels are especially well-suited for eliminating *random noise* such as quantum mottle, whereas frequency processing (specifically, band-pass filtering) is more effective for suppressing *periodic noise* such as electronic noise.

Preparation for Display

All images are stored by a PAC system in the standardized DICOM format so that it is generally compatible with various acquisition, storage and display systems. But, when an image comes up for display on a particular monitor screen, the number of pixels in the matrix of the image must match the number of pixels available on the display monitor so that the image "fits" on the screen. For example, the digital image stored by the computer might consist of a matrix 1000 pixels tall by 800 wide, whereas the hardware pixels of the display monitor itself compose a matrix 100 pixels tall by 667 pixel wide (an *aspect ratio* of 3:2). To avoid inadvertent "cropping" of the image, possibly clipping off anatomy, either the computer or monitor must calculate how to best fit this image to the screen.

In addition to this geometrical or spatial compatibility, the dynamic range being made available by the computer must match the bit depth of the display monitor—otherwise, windowing at the monitor could result in data clipping (see Chapter 7) or other issues. Typically, the capabilities of the display monitor are less than those of the computer in every respect. Algorithms to assure that all of these parameters align are usually included within the display monitor itself.

Chapter Review Questions

1. Detail processing is defined by its ability to treat the fine details of an image _____ from the larger, gross structures in the image.

2. Detail processing is able to *decouple* local contrast from _____ contrast.

3. On the display monitor, if each row of the hardware matrix has 600 pixels across it from left-to-right, the inherent lateral *frequency* of the monitor is _____ Hertz.

4. In the complex waveform that represents one row of an actual displayed image, the amplitude or height of each pulse represents the displayed _____ on the monitor.

5. In the complex waveform that represents one row of an actual displayed image, a very wide wave pulse indicates a large structure that occupies a large number of _____ in the row.

6. A complex wave is formed when wave pulses of different frequencies are _____ over each other.

7. On a waveform graph, adding a small *negative* pulse to a large positive pulse results in a large pulse with a _____ in it.

8. The various frequencies of the displayed image can be separated by the mathematical process called _____ _____.

9. *Multiscale* processing separates the original image into _____ or more frequency layers.

10. The *background density* of an extremity image is not seen on high-frequency image layers because it is treated as a _____ "structure" by frequency processing.

11. After targeted frequency layers of the image are treated, all the layers are compiled (added) back together using _____ *Fourier transformation.*

12. In frequency *band-pass filtering,* a selected _____ of the image is left out upon reconstruction.

13. *Low-pass filtering* removes _____-frequency layers from the image.

14. *High-pass filtering* and *background suppression* both result in edge _____.

15. While frequency operations are applied, a log of the _____ of each object in the image is kept so that the final image can be reconstructed in its spatial matrix.

16. A *kernel* can also be used for detail processing when simple _____ factors are placed in its cells. The products from these calculations are added together to insert into the centered pixel.

17. For detail processing with kernels, whether *smoothing, edge-enhancement* or *background suppression* occurs depends upon the _____ between the factors entered in its cells.

18. When used for filtering, the *larger* a kernel is, the _____ the structures it will suppress in the image.

19. When used for filtering, too small a kernel may remove important small _____ from the image.

20. When a kernel is used to separate a "gross-structure" layer of the image, then a reversed *positive* copy of that layer is subtracted from the original image, what is this procedure called?

21. What two aspects of the final digital image must be formatted to be compatible with the display monitor?

Chapter 9

MANIPULATING THE DIGITAL IMAGE: OPERATOR ADJUSTMENTS

Objectives

Upon completion of this chapter, you should be able to:

1. Describe the advantages and limitations of employing alternate procedural algorithms.
2. Clarify the terminology and effects of window level and window width in relation to the final displayed image and to each other.
3. Explain the applications of *smoothing, edge enhancement, background suppression, targeted area brightness correction* and *image reversal.*
4. Explain the limitations of *smoothing, edge enhancement, background suppression, targeted area brightness correction* and *image reversal.*
5. State the applications and legal implications of *dark masking.*
6. Describe *image stitching, dual-energy subtraction, and grid-line suppression* software.

Operator adjustments cover a wide selection of features that can be applied to the image by the radiographer or radiologist after the initial image is displayed. *Leveling and windowing* are routinely applied on a daily basis, and are actually adjustments to the *gradation* processing already completed. A number of proprietary features from each manufacturer can be applied at the push of a button, and include such examples as applying a "*soft* or *hard look*" to the image on GE equipment or applying *enhanced visualization processing (EVP)* on CareStream equipment to bring out small details. Additional features for specialized equipment include *image stitching* that pieces together multiple views from a scoliosis series or *Grid line suppression* software. All of these operator adjustments fall under the category of *postprocessing* as we have defined it.

Processing Algorithms

At the x-ray machine console, when the radiographer selects a radiographic procedure that is about to be performed, a digital processing program tailored to that anatomy is being selected. This program includes the type of reference histogram that the incoming histogram will be compared to for histogram analysis (see Chapter 5), the permanent LUT that will be used for re-scaling (see Chapter 6), the anatomical LUTs that will be used for default gradation processing (see Chapter 7), and all the default detail processing features that will be applied before the initial image is displayed (see Chapter 8).

After the advent of computed radiography, it did not take long for radiographers to discover by experimenting at the console that the appearance of the displayed final image could be changed by selecting an "alternate procedure algorithm" for processing. An example is the PA chest projection in Figure 9-1 where chest radi-

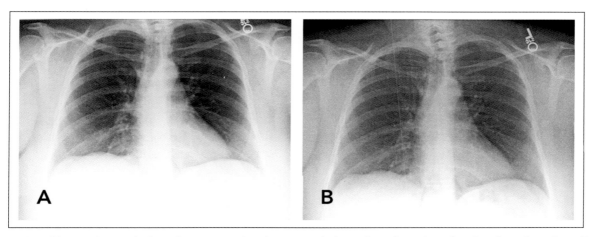

Figure 9-1. Chest radiograph digitally processed with a normal chest algorithm, *A*, and with a foot algorithm, *B*. In this case, the foot algorithm results in increased gray scale and an increased number of visible radiographic details. (From Q. B. Carroll, *Radiography in the Digital Age,* 3rd ed. Springfield, IL: Charles C Thomas, Publisher, Ltd., 2018. Reprinted by permission.)

ograph *B* was entered into the computer as a "foot" procedure. You can see that this resulted in a chest radiograph with considerably lengthened gray scale, compared to the normal "chest" algorithm used for radiograph *A* on the same patient. To rule out some diseases of the chest, radiograph *B* may serve better than *A*.

In order to reprocess an image with an alternate anatomical algorithm (for most manufacturers), touch the histogram graph displayed at the console while the image is on display. A new screen will appear with a list of body parts or procedures, from which you may select the alternate algorithm to apply.

The great majority of digitally processed radiographs should have ideal quality using the default processing settings. Therefore, the use of alternate algorithms should be the exception rather than the rule. Discretion should be used. There should be a demonstrable benefit in the altered image, and a clear rationale for making the change. Most importantly, a supervisor or radiologist should approve of using an alternate algorithm as a regular practice for any procedure.

Although the use of an alternate algorithm can be diagnostically beneficial, as shown in Figure 9-1, it has ramifications for image storage, record-keeping, and legal liability. First, remember that some changes will result in a narrower data set for the image that can limit the radiologist's ability to window the image to his/her satisfaction. If the altered image is "saved" or sent into the PAC system as a *replacement* for the original file, the data set from the original image is permanently lost and cannot be retrieved. The best policy is to first make a *copy* of the original image, reprocess only this copy under the alternate algorithm (preferably with annotation explaining the change), and then save the copy into the PAC system as a supplemental image in the series.

The altered image is also saved into the PAC system under a different DICOM header as a different *procedure* for this patient, which can cause complications in record-keeping and retrieval of radiographic studies, and creates a potential for legal problems. Lawsuits have been decided on the basis that the information in a *saved* radiographic image file had been tampered with. Imaging departments should develop clear policies for saving altered images into the PAC system.

Windowing

The brightness (density) and the contrast (gray scale) of an image can both be adjusted at the console *as the image is being viewed.* Col-

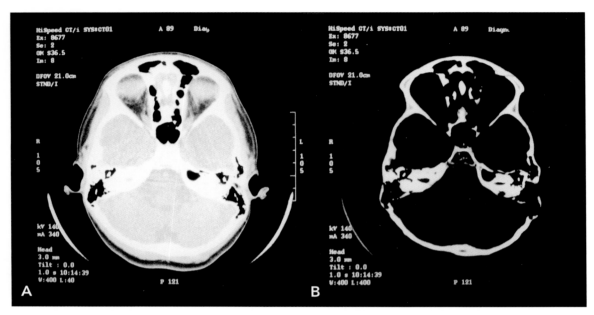

Figure 9-2. CT scans of the head show, from **A** to **B**, that an increase in window level (L) from 40 to 400 has the effect of darkening the overall image. (Window width was fixed at 400 for both images.) (From Q. B. Carroll, *Radiography in the Digital Age,* 3rd ed. Springfield, IL: Charles C Thomas, Publisher, Ltd., 2018. Reprinted by permission.)

lectively, these adjustments are generally referred to as *windowing.* The generic terms best suited to these two changes have been long established by the oldest computerized medical imaging modality, CT. The CT technologist refers to them as adjusting the *window level (WL)* and adjusting the *window width (WW).*

Window level corresponds to the average or overall *brightness* of the displayed image. At the console, other labels used by some manufacturers for window level include "brightness," "center," or "density," (and for Fuji, the "S" number), all of which express the same general concept. In computerized tomography (CT), increasing the window level value to a higher number always corresponds to making the image darker Figure 9-2. This correlates with increasing density, but where manufacturers use the term *brightness,* the numerical scale will be inverted, so increasing the brightness value would correspond to decreasing the window level. *Window level is opposite to brightness.*

Shown in Figure 9-3, as the window level is raised or lowered, the entire gray scale of the

image moves with it. In this figure, the *dynamic range* of available densities (or pixel values) is depicted at the left. Note that in **B**, when the window level is increased, the *gray scale selected from the dynamic range does not change,* that is, the gray scale for image **A** is 5 densities displayed, and the gray scale for image **B** is still 5 densities displayed. What has changed is that the *center-point* for the 5 densities, indicated by the dashed arrow, has been moved from a light gray to a much darker gray density. We can define this center-point as the "average" density of the displayed gray scale, and conclude that increasing the window level has made the image darker *overall.*

Figure 9-4 graphically illustrates an increase in *window width* from **A** to **B**. Note that the *center-point* of the gray scale is maintained at step 6, indicating that *average* brightness or overall density is unchanged. But, for image **B**, with a (vertically) wider window, there are now nine different shades of gray displayed in the image rather than 5. Gray scale has been lengthened. In CT, increasing the window width value to a higher number always corresponds to lengthening the

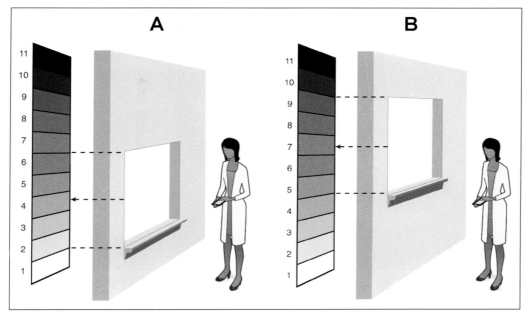

Figure 9-3. From *A* to *B*, when the window level is raised, the overall image is darkened. However, the range of gray shades (window width) remains unchanged at 5 steps. The window level value is the average darkness or center of the gray scale, represented by the arrow. (From Q. B. Carroll, *Radiography in the Digital Age,* 3rd ed. Springfield, IL: Charles C Thomas, Publisher, Ltd., 2018. Reprinted by permission.)

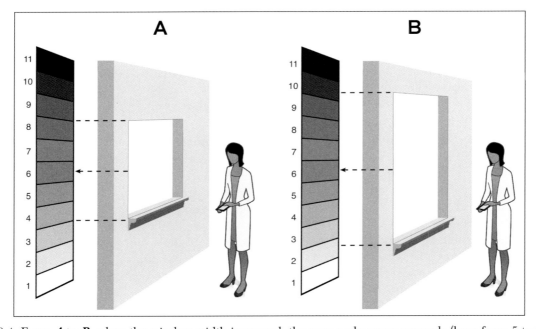

Figure 9-4. From *A* to *B*, when the window width increased, the gray scale range expands (here from 5 to 9 steps). This can be done without changing the average darkness (window level) represented by the arrow at the center of the gray scale. (From Q. B. Carroll, *Radiography in the Digital Age,* 3rd ed. Springfield, IL: Charles C Thomas, Publisher, Ltd., 2018. Reprinted by permission.)

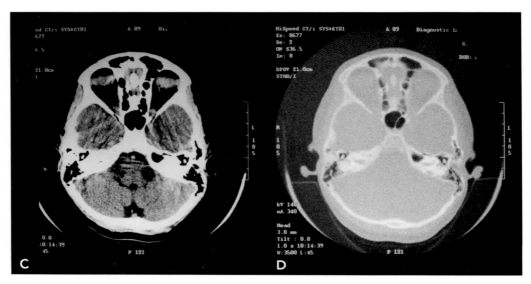

Figure 9-5. CT scans of the head show, from **C** to **D**, that an increase in window width (W) from 97 to 3500 has the effect of expanding the gray scale, (but with equal overall brightness). (From Q. B. Carroll, *Radiography in the Digital Age,* 3rd ed. Springfield, IL: Charles C Thomas, Publisher, Ltd., 2018. Reprinted by permission.)

gray scale (Figure 9-5). Where manufacturers use the label *contrast* at the console, the numerical scale will be inverted, so increasing contrast would correspond to decreasing the window width. *Window width is opposite to contrast.*

(For Fuji equipment, the "L" number corresponds to "latitude" or the extent of the gray scale. To lengthen gray scale, turn this number up. To increase contrast, turn the "L" number down.)

It is important to emphasize that as demonstrated in Figure 9-3 (also see Figure 3-2 in Chapter 3), overall image brightness or density can be changed *without changing gray scale.* Conversely, in Figure 9-4, image contrast or gray scale can also be changed *without altering overall brightness or average density.* Restated, WL can be changed without changing WW, and WW can be changed without altering WL. They are independent of each other. In the latent image at the IR, some variables such as scatter radiation can alter *both* density and contrast, but this does not mean that there is a *causative* relationship between these two image qualities: An image can be made darker or lighter without altering the contrast. Likewise, the contrast or gray scale of an image can be changed without moving the *average* or *overall* brightness.

In daily practice, adjusting the window level is frequently expressed in shortened form as "levelling" the image, and adjusting the window width is often shortened to "windowing" the image. In this abbreviated parlance, levelling is adjusting the brightness of the image, windowing is adjusting the contrast or gray scale.

Postprocessing Features

Smoothing

Smoothing algorithms remove some of the smallest details (highest-frequency layers) from the image. There are several results: First, the *edges* of bony structures appear "softened" due to a slight reduction in local contrast. This is where *smoothing* gets its name, as the "hard edge" is taken off of these bony structures. Second, *noise is reduced* in the image by mathematical *interpolation* which averages the pixel value of small white or black specks with their surrounding pixels, making them "blend in." Third, this same interpolation (averaging) process corrects for dead or stuck dexels in the image receptor, discussed in Chapter 5. Dead or stuck dexels constitute a form of noise, so anything that reduces

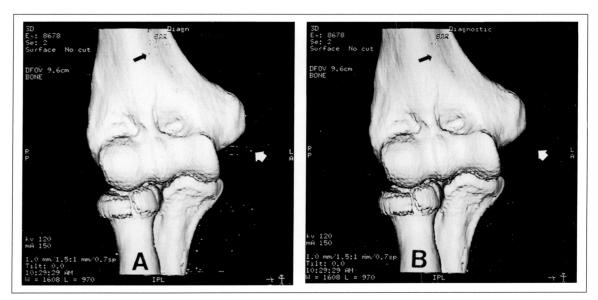

Figure 9-6. 3D CT scans of the elbow demonstrate the effects of smoothing in image *B*, where noise artifacts over the bone (*black arrow*) and the background (*white arrow*) have been removed, and detail edges have less contrast. (From Q. B. Carroll, *Radiography in the Digital Age,* 3rd ed. Springfield, IL: Charles C Thomas, Publisher, Ltd., 2018. Reprinted by permission.)

noise will also suppress the effect of dead or stuck dexels. Manufacturers may use different proprietary names for this feature, but all manufacturers make smoothing available in some form, and most also employ it as one step in their default processing. An example of smoothing is provided in Figure 9-6.

Smoothing is recommended to correct for moderate amounts of *mottle* appearing in the image. However, bear in mind that *severe* mottle indicates gross underexposure to the IR for the initial projection. In Chapter 4, we emphasized that severe underexposure results in a true loss of information within the data set that is fed into the computer. *This lost information cannot be recovered using smoothing or any other processing feature.* Radiographs presenting severe mottle from underexposure should be repeated.

It can generally be assumed that *the application of any digital processing feature will have a trade-off for the advantage gained, that is, "there is always a price to pay."* Consider a radiographic image that already presents somewhat low overall contrast: By reducing local contrast, smoothing can have the negative effect of losing some of the

small details in such an image, details that constitute useful diagnostic information. This is because the local-contrast effect of smoothing is *added* to the previous loss of overall contrast. (We have stated that changing local contrast, at the scale of only small details, does not visibly impact global contrast; However, changing global contrast *does* affect local contrast because small details are included.)

Edge Enhancement

Edge enhancement is used when the radiologist needs to better visualize small details at the edges of bones, cartilage, or organs such as the kidneys. It is simply an increase in local contrast that can make small pathological changes, such as a hairline fracture, more visible to the human eye. Figure 9-7*A* is an example of an edge-enhanced image of the lateral C-spine, (compare to the default image *B*). Image *C* is an example of applied smoothing. Collectively, these images in Figure 9-7 illustrate the "look" feature used by General Electric, in which a "hard" look is actually an edge-enhanced image and a "soft" look

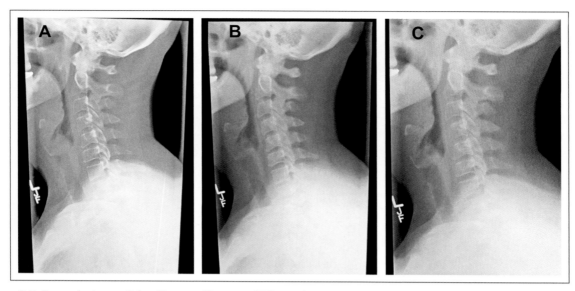

Figure 9-7. Lateral views of the C-spine illustrate GE's application of a "hard look," **A**, where edge enhancement might better demonstrate a hairline fracture, normal default processing in **B**, and a "soft look," **C**, where smoothing has been applied. (From Q. B. Carroll, *Radiography in the Digital Age,* 3rd ed. Springfield, IL: Charles C Thomas, Publisher, Ltd., 2018. Reprinted by permission.)

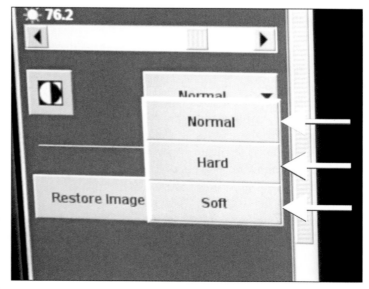

Figure 9-8. Selections from the "look" feature on a DR unit manufactured by GE include *normal, hard,* and *soft.* The "hard" look setting applies edge enhancement. The "soft" look applies a smoothing algorithm. (From Q. B. Carroll, *Radiography in the Digital Age,* 3rd ed. Springfield, IL: Charles C Thomas, Publisher, Ltd., 2018. Reprinted by permission.)

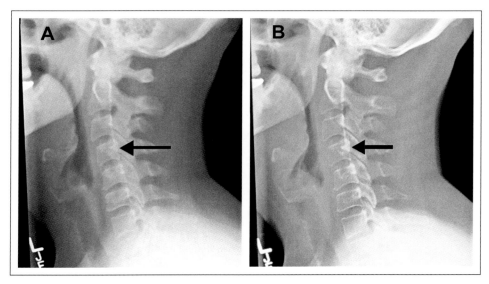

Figure 9-9. In image **B**, when the manufacturer's *default* edge enhancement feature was applied to a lateral C-spine image already possessing high contrast, the radiologist noted that a hairline fracture could be *obscured* in the light triangular area indicated by the arrow. The default setting was too high for edge enhancement. (From Q. B. Carroll, *Radiography in the Digital Age,* 3rd ed. Springfield, IL: Charles C Thomas, Publisher, Ltd., 2018. Reprinted by permission.)

is actually a smoothed image. Figure 9-8 shows the GE menu screen to select these features.

It is possible to *over-enhance* local contrast. The manufacturer can have the default setting for edge enhancement set too high, even to the extent of losing details as shown in Figure 9-9. Especially if an image already possesses high global contrast, applying edge-enhancement can potentially damage the visibility of important details, affecting diagnosis.

The trade-off for using edge enhancement is that it also enhances image noise, which may not have originally been at a bothersome visible level but now is. The level of *visible* noise in the displayed image is increased. Certain image artifacts can also be created, such as the *halo effect,* in which, for the boundary between two anatomical structures, the darker density side of the edge is over-darkened while the lighter side is lightened even more. Figure 9-10 demonstrates halo artifacts.

It is important to realize that on most digital units, the degree to which either smoothing or edge enhancement are applied *can be customized* by the quality control technologist by accessing password-protected menus such as the one shown in

Figure 9-11. In consultation with the radiologists, if it is determined that the manufacturer's original default settings apply these features too extremely for any particular procedure, they can be adjusted to a milder application or vice versa.

To summarize, smoothing can make details more visible by *reducing the noise* that superimposes them. Edge enhancement can make details more visible by *increasing their local contrast.* In both cases, the resulting image should always be carefully compared with the original to ensure that there was a diagnostic improvement in image quality and not a negative impact on it.

Background Suppression

Background suppression algorithms reduce the contrast only of larger mid-frequency and low-frequency structures. As one might expect, the result is very similar to edge enhancement, but background suppression can be more effective for specific situations like a muscular fat pad overlapping bone marrow details.

Manufacturers use different proprietary names and formats for these features. The three

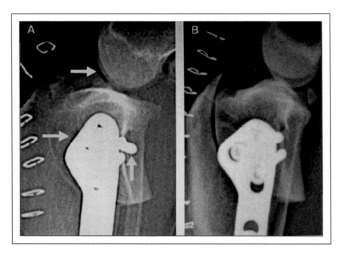

Figure 9-10. A lateral view of an animal's knee shows the dark halo artifact around the femoral condyles and the orthopedic plate (*left*) with edge enhancement engaged. In ***B***, these artifacts were eliminated by simply turning off the edge-enhancement feature. (From Q. B. Carroll, *Radiography in the Digital Age,* 3rd ed. Springfield, IL: Charles C Thomas, Publisher, Ltd., 2018. Reprinted by permission.)

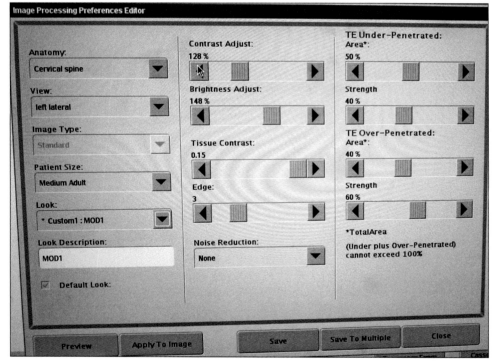

Figure 9-11. Customization menu screen for the "look" feature on a DR unit manufactured by GE. In addition to general contrast, local contrast, and brightness for a particular procedure, the lower mid-section includes adjustments for edge enhancement ("edge") and smoothing ("noise reduction"). The right section adjusts specified areas of the image for darkening or lightening, replacing the need for compensating filters in many instances and correcting for the underexposed shoulder area on lateral C-spine projections (see Figure 9-12). (From Q. B. Carroll, *Radiography in the Digital Age,* 3rd ed. Springfield, IL: Charles C Thomas, Publisher, Ltd., 2018. Reprinted by permission.)

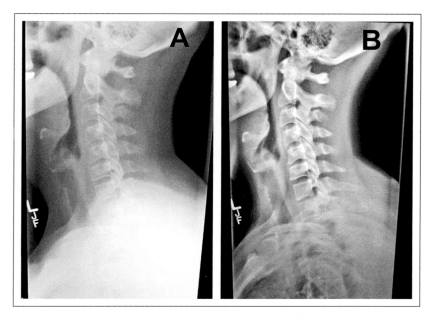

Figure 9-12. Targeted area brightness correction applied to the lateral C-spine view, **B**, to demonstrate the cervicothoracic vertebrae (C7-T2) that were originally obscured by superimposed shoulder tissues, **A**. Here, General Electric's "underpenetrated" function was used to identify the lower 1/3 (30%) of the image and apply gradation processing adjustments to darken only that portion of the image. (From Q. B. Carroll, *Radiography in the Digital Age,* 3rd ed. Springfield, IL: Charles C Thomas, Publisher, Ltd., 2018. Reprinted by permission.)

generic features described here (smoothing, edge enhancement, and background suppression) are available in one form or another on nearly all brands of digital equipment.

Targeted Area Brightness Correction

For decades, a bane of radiographers has been the lateral cervical spine projection which was often too light in the C7–T2 area where the patient's shoulders overlapped the spine, Figure 9-12***A***. Several manufacturers now include software programs that *target specific portions of the image for brightness correction.* Image ***B*** in Figure 9-12 demonstrates the correction applied to the lower 1/3 (30%) of the image.

In Figure 9-11, the menu for this feature on a GE unit is seen in the right-hand portion of the screen as "TE Under-Penetrated" and "TE Over-Penetrated" where "TE" stands for "tissue equalization." Note that for either application, there is a setting for the percentage "area" of the image to be affected, and a setting for the "strength" or

degree of the application to be applied. The proprietary nomenclature used by different manufacturers can be confusing for the student. On this GE menu, the terms used seem to imply that "penetration" is being adjusted. Of course, we cannot go back in time and change the actual penetration of the x-ray beam during the exposure. These settings actually use gradation processing to adjust the brightness of the image in areas the radiographer considers to be "under" or "over-penetrated" in appearance.

Other Postprocessing Features

Image reversal, sometimes called "black bone," is demonstrated in Figure 9-13. All of the pixel values within the image are simply changed from high to low numbers and vice versa. This results in "positive" black-on-white" image rather than the standard "negative" white-on-black image used for conventional radiographs. Radiologists occasionally use this feature to *subjectively* bring out details they are trying to see. Image

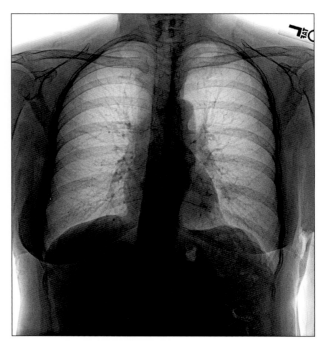

Figure 9-13. Image reversal or "black bone" changes all pixel values from positive to negative and vice versa. No new details are present, but subjectively some pathologies may be easier to see for some individuals. (Courtesy, Robert Grossman, RT.)

reversal can be a useful and productive application, but it is important to remember that no new details or information are being *created* in the image. In fact, contrast remains completely unchanged as well, so any "improvement" in the image is purely subjective and a matter of personal preference for different diagnosticians.

Dark masking refers to the placement of a black border around and image, as opposed to a white one. Conventional film radiographs were hung on "viewbox" illuminators. If the film was smaller than the viewbox, there was considerable glare from the light surrounding the image. Any kind of excessive ambient light around a radiograph will *visually* impair the contrast apparent in the image. To prevent this, the entire display screen around a radiograph should be black. Before a digital image is sent into the PAC system, dark masking should be applied if there is any white border around it. This improves the apparent visual contrast and is generally recommended to be applied to all images.

After dark masking became available, radiographers quickly realized they could use it to effectively "re-collimate" an image which had been taken with too large an exposure field. Attempting to hide poor collimation during the initial exposure is not only unethical, but also has legal ramifications that have been decisive in settling malpractice lawsuits: The radiologist is held legally responsible for *all diagnostic information present in a radiograph.* For example, a chest procedure may have been ordered only to rule out an enlarged heart, but if subcutaneous pneumothorax (air pockets under the skin) is also present from disease or trauma, the radiologist *must* report these findings even if they are unrelated to the purpose for which a clinician ordered the procedure. If a radiograph uses dark masking to effectively collimate the image more tightly, and some of this pathology under the skin is thus removed from the image, the radiologist, hospital or clinic, and radiographer are legally responsible for the medical consequences. Radiographers must properly collimate the x-ray beam for every projection. If this is not done during the initial exposure, it is neither legal nor ethical to try to hide it after the fact.

Image stitching is a great invention that replaces the need for the heavy, oversized cassettes that used to be used for scoliosis series. In order to fit the entire spine on one exposure, cassettes that were up to 106 cm (42 inches) long had to be used. With computerized technology, three separate CR exposures can be quickly taken, cervicothoracic, thoracolumbar, and lumbosacral, to cover the entire spine while the patient remains in position. The three projections overlap each other by only a few centimeters. Computer software then uses an alignment "grid" to perfectly superimpose these images, crop and combine them to form a compound image of the entire spine that is presented at the display monitor. DR systems use the same type of "stitching" software, but typically only require 2 exposures which can be taken much quicker by moving the DR detector rather than changing CR cassettes.

Dual-energy subtraction can separate the original image into a *tissue only image* and a *bone only*

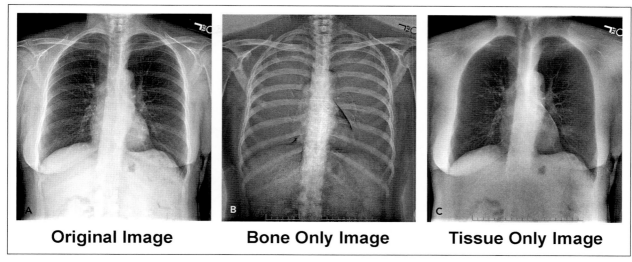

Original Image **Bone Only Image** **Tissue Only Image**

Figure 9-14. Dual-energy subtraction can separate an original image, *A*, into a bone-only image, *B*, and a tissue-only image, *C*. In the soft-tissue image *C* a circular lesion is seen in the left lung to the side of the apex of the heart, which is largely obscured by the 9th rib in the original image *A*. (Courtesy, J. Anthony Siebert, Ph.D. Reprinted by permission.)

image, as shown in Figure 9-14. This allows us to determine whether a pathological lesion is in the soft tissue just behind or in front of a bone or within the bone itself. This is especially applicable in the chest where the thin ribs overlap the lung tissues both anteriorly and posteriorly. To separate the soft tissues from the bony skeleton, it is first necessary to obtain a *high energy image* and a *low energy image* (Figure 9-15). This can be done by changing either the kVp or the amount of filtration used in the x-ray beam.

For the variable kVp method, after a typical high-kVp (120 kVp) chest projection is exposed, the x-ray machine must be capable of quickly and automatically switching to a low-kVp technique and re-expose the detector plate. Typically, it takes at least 200 milliseconds for the machine to switch the kVp and be exposure-ready, and the patient must not move or breath during all the time it takes to acquire the two exposures. Increasing the kVp raises the *average* energy of the x-ray beam so it is more penetrating and much fewer photoelectric interactions are taking place within the patient.

The variable filtration method has the advantage of requiring only one exposure, so that patient motion is less probable. Here, we employ a *multilayer cassette,* two detector plates with a thick x-ray filter placed between them (Figure 9-16). The front plate records the "low-energy" image using the typical x-ray beam. Behind this, the filter "hardens" the remaining x-ray beam before it exposes the back plate. Removing low-energy photons raises the *average* photon energy. This has the same effect as increasing the kVp: The beam becomes more penetrating, and many fewer photoelectric interactions take place.

Now, with either method, the *key* is that photoelectric "drop-off" is much more dramatic for soft tissue than it is for bone tissue, as shown in Figure 9-17. At higher *average* kVp levels, photoelectric interactions drop precipitously for soft tissue. In fact, photoelectric interactions almost never happen when the incoming photon's energy is above 60 kV (Figure 9-17). In bone, however, photoelectric interactions are still occurring at average kV levels above 100, and photoelectric drop-off follows a shallower slope. This allows the computer to discriminate between soft tissue and bones. The computer compares the two images, high-energy and low-energy, and can identify those areas where x-ray absorption dropped off more dramatically as soft-tissue areas. It can then reconstruct one image using only the

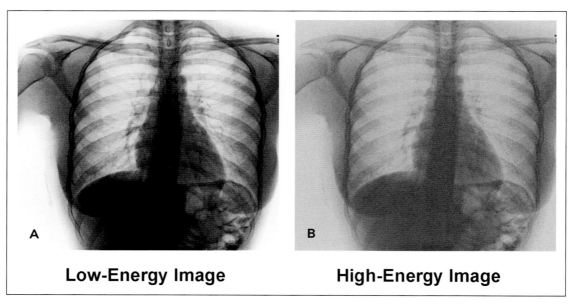

Low-Energy Image　　　**High-Energy Image**

Figure 9-15. Positive (reversed) images used for dual-energy subtraction. *A* is the low-kVp or non-filtered image, *B* is the high-kVp or filtered image. The difference in the drop-off of photoelectric interactions (See Figure 9-17) allows the computer to distinguish between bone tissues and soft tissues to select for subtraction. (Courtesy, J. Anthony Siebert, Ph.D. Reprinted by permission.)

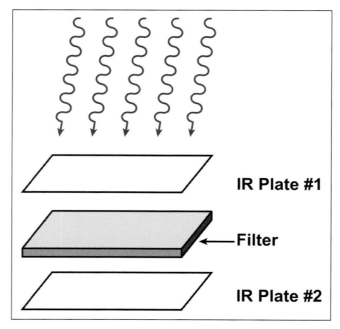

Figure 9-16. Filtration method for dual-energy subtraction. IR Plate #1 is used to produce the "unfiltered" or low-energy image, IR Plate #2, behind the metal filter, produces the filtered (high-energy) image. (From Q. B. Carroll, *Radiography in the Digital Age,* 3rd ed. Springfield, IL: Charles C Thomas, Publisher, Ltd., 2018. Reprinted by permission.)

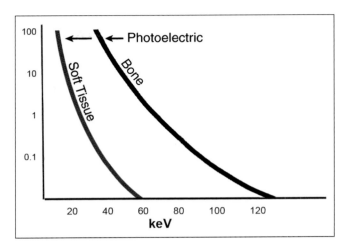

Figure 9-17. At higher energy (keV) levels, the photoelectric interaction drops off more precipitously for soft tissue than for bone. Therefore, the contrast for soft tissue changes more dramatically, providing a measurement the computer can use to distinguish between soft tissue and bone to select for subtraction.

data for soft tissue, and a separate image using only the data for bone. Figure 9-14 illustrates the impressive result and advantage for diagnosis.

Grid line suppression software was developed to allow the use of *stationary* grids in mobile and trauma radiography, and then "erase" the grid lines they cause with digital processing. *Tomographic artifact suppression* uses a nearly identical process to remove "streak" artifacts caused by linear tomography equipment. In both cases, frequency processing is able to identify these long lines in the image as extremely "large" structures with a very low frequency *in a single axis* (e.g., lengthwise). We simply filter the image layer which contains structures of only 1 to 3 hertz in the lengthwise axis. However, *virtual grid software,* discussed in Chapter 4, is trending to replace the use of grids altogether, making grid line suppression superfluous.

These are just a few examples of the most common postprocessing features available from different manufacturers of x-ray equipment. Each manufacturer typically offers additional unique features or unique combinations of algorithms to achieve different effects.

Chapter Review Questions

1. Because of its ramifications for record-keeping and legal liability, the use of alternate anatomical algorithms to process images should be the _____ rather than the rule.
2. An image processed under an alternate algorithm is saved into the PAC system under a different DICOM header as a _____ procedure than what was ordered.
3. Generally, any altered image should not be saved into the PAC system as a _____ for the original image.
4. At the console, adjustments to the brightness and contrast of the displayed image as it is being viewed are collectively referred to as _____.
5. *Brightness, center, density* and *S number* are all terms that refer to setting the window _____.
6. A very bright image is set to a _____ window level.
7. To lengthen the gray scale of an image, _____ the window width.
8. For the displayed image, can the window level be adjusted without affecting image contrast?

9. In practical use, "windowing" the image may be a shortened expression for adjusting the window _____.

10. Which common postprocessing feature "softens" the appearance of the displayed image, reduces mottle, *and* corrects for stuck or dead pixels?

11. *Severe* mottle implies severe underexposure and a loss of _____ which cannot be recovered by digital processing.

12. The local contrast of small details can be changed without affecting the overall, global contrast of the displayed image, but changing global contrast _____ affect local detail contrast.

13. Image _____ can be lost either by applying smoothing to an image already possessing low contrast, or by applying edge-enhancement to an image already possessing high contrast.

14. The negative trade-off for applying edge enhancement is that it also enhances image _____.

15. Background suppression has similar effects to _____ enhancement.

16. For the lateral C-spine view, cervicothoracic vertebrae superimposed by shoulder tissue can be better demonstrated by using *targeted area _____ correction.*

17. *Image reversal* results in a _____ image, defined as "black-on-white" in appearance.

18. Although image reversal may *subjectively* bring out details to the observer, no new _____ is added to the original image.

19. Dark masking is generally recommended because ambient light around a radiographic image visually reduces its _____.

20. Dark masking should not be used to effectively "recollimated" an image originally taken with too large an exposure field, because radiologists are held legally liable for _____ diagnostically useful information on a radiograph.

21. Scoliosis radiographs surveying the entire spine can be combined into a single image using image _____ software.

22. Where subtle soft-tissue lesions overlap bony structures such as the ribs, making them difficult to see, the original image can be separated into a *tissue-only* image and a *bone-only image* using dual-energy _____.

23. What are the two methods for obtaining a low-energy image and a high-energy image?

24. Grid-line suppression software identifies grid lines as _____-frequency structures in a single axis.

Chapter 10

MONITORING AND CONTROLLING EXPOSURE

■ ■ ■ ■ ■ ■ ■ ■ ■ ■ ■ ■ ■ ■ ■ ■ ■ ■ ■

Objectives

Upon completion of this chapter, you should be able to:

1. Explain the implications of digital processing speed class for patient dose and image mottle.
2. Define *exposure indicator (EI)* and *target EI* and their relationship to speed class.
3. Interpret the percentages of actual exposure deviation for *deviation index (DI)* readouts of plus- or minus- 0.5, 1.0, and 3.0.
4. Recite by memory the meaning and the recommended action for the *DI* ranges –0.5 to +0.5, +1 to +3, greater than +3, –1 to –3, and lower than –3.
5. List common potential causes of a corrupted *EI* leading to an erroneous *DI* readout.
6. Distinguish between overexposure and *saturation*, their appearance, and their causes.
7. Identify the limitations of the *deviation index* in relation to accurately evaluating both patient dose and radiographic image quality.
8. Describe how the advent of digital imaging has shifted the concept of "controlling factors" for image quality from radiographic technique to digital processing. Distinguish between factors controlling the qualities of the latent image reaching the IR and those of the *final digital image* displayed at the monitor.

Speed Class

The *speed* of any imaging system expresses its sensitivity to radiation exposure. Historically, for film-based x-ray systems, we simply chose a mid-level target density (medium gray) to be produced on a film and compared the amount of exposure different types of film required to produce that density. If film brand *A* required twice as much exposure as brand *B*, we said that film *A* had *one-half* the speed. The sensitivity of different types of film could be compared this way, just as is done in photography. The standard speed value was set at *100*. Thus, a 50-speed film was twice as "slow" and required twice the exposure.

Each type of image receptor for modern digital x-ray equipment also has an *inherent* speed that is still expressed using the old system of notation. For example, a typical CR phosphor plate has a speed of about 200, twice as sensitive as conventional 100-speed x-ray film. However, for a modern digital x-ray unit, this is only half of the story—its speed is based primarily upon the *digital processing* used rather than the IR alone. Since the processing can be set to deliver a darker or lighter image, the *inherent speed of the IR is important only at the image acquisition stage*. It is a novel aspect of digital equipment that *the speed at which it operates can be selected without any physical change to the IR*. When a new digital machine is installed, the department manager and quality control supervisor can tell the manufacturer's technical representative what speed they want, and installers can set the equipment to operate at this speed as a default setting. Because this setting is not an inherent physical attribute of the IR (or of the machine itself), and can be changed, we refer to it as the *speed class* rather than just the "speed."

Operation at a speed class of 100 assumes an average exposure of 20 microgray penetrating through the patient to reach the IR. If we operate the machine at a speed class of 200, we assume that remnant x-ray beam exposure to the IR will be at least 10 microgray, at 400 speed, 5 microgray. Insufficient exposure can lead to a mottled image, and manufacturers have installed many CR machines with the default speed set to 200 in an effort to double-ensure against the appearance of mottle. But, *nearly all DR and CR machines can be operated at a speed class of 350-400, without the appearance of substantial mottle. Doing so saves patient exposure to harmful radiation.* As described in Chapter 4, very slight mottle can be acceptable to the radiologist and, in fact, is an indication that the lowest practicable technique has been used in the interest of the patient.

A Brief History of Exposure Indicators

We have described how with the compensating power of rescaling and other digital processing features, the image almost always turns out right. For film-based radiography, there was immediate feedback as to whether too much or too little technique was used, in the form of dark or light images. In the digital age, this immediate feedback is missing, (the one exception being the appearance of extreme mottle, which would indicate gross *underexposure*. For *overexposure*, there is simply no visual cue.)

Therefore, manufacturers developed various types of exposure indicators that could be displayed with each image as a numerical value on the monitor screen. A very high exposure indicator (EI) would indicate that there was an excessive amount of x-ray exposure *at the IR*. A very low EI would indicate that the image receptor received an insufficient amount of exposure.

Each manufacturer designated a *target (EIT)* that represented the "ideal" amount of x-ray exposure at the IR. The EIT is *not* inherent to the x-ray unit or processor. Managers, in consultation with radiologists, physicists and the manufacturer, establish the desired image quality and patient dose outcomes for a department. The processing *speed class* for different procedures is thus determined, and an appropriate EIT is derived for that speed class.

For most manufacturers, the EI is *derived* from the mid-point of the image histogram before any changes are made to it by rescaling. In other words, the *average pixel value* from the histogram is used to extrapolate what the exposure in microgray must have been at the IR. *This is not a direct measurement of actual x-ray exposure to the patient, or even to the IR, but an estimate based on pixel values in the "raw" histogram.*

All kinds of mathematical approaches are used by more than 14 different manufacturers to calculate an exposure indicator. They use different formulas, different names, and different formats for presentation on the monitor screen, an intolerable situation. A sampling of the different names used for exposure indicators includes *Log of Median (Agfa), Exposure Index (CareStream, Philips, and Siemens), S Value (Fuji), Reached Exposure Value (Canon and Shimadzu), Detector Exposure Index (GE), Sensitivity Number (Konica),* and *Dose Indicator (Swissray).* Among all these manufacturers, three general types of scales were used to report their exposure indicators—some used logarithmic scales, some proportional scales, and amazingly, some even used *inversely* proportional scales such that the lower the reported indicator value, the higher the exposure!

The Deviation Index: Acceptable Parameters for Exposure

Reason eventually prevailed in 2009 when the American Association of Physicists in Medicine (AAPM) finally concluded that, "A standardized indicator . . . that is consistent from manufacturer to manufacturer and model to model is needed." In their report, *An Exposure Indicator for Digital Radiography,* Task Group 116 of the AAPM proposed a standardized *deviation index (DI)* that can be used by all manufacturers, regardless of their specific method for calculating an exposure indicator. They recommended that the DI "should be prominently displayed to the operator of the digital radiography system immediately after every exposure." The DI readout and an indicator of the actual exposure delivered to the image detector in microgray units should be included in the DICOM header for every image. (The DICOM header is a "metadata" record for each image that can be accessed in the PAC system—see Chapter 13.)

Most manufacturers are now installing the generic *deviation index* readout for all newly manufactured x-ray equipment, and many are retrofitting older units with it. It is important to understand that although the format and criteria for the *displaying* the deviation index has now been standardized, it is derived from the exposure indicator, and each manufacturer is still allowed to use their disparate methods and formulas to calculate their EI. Most exposure indicators are derived from the "raw" histogram. And, as shown in Chapter 5, the histogram is subject to skewing from exposure field recogni-

tion errors, segmentation failure, the presence of lead shielding, large radiopaque prostheses or orthopedic hardware within the field, extremes in collimation, scatter radiation, and other unusual circumstances. *Anything that can cause errors in histogram analysis can lead to a corrupted exposure indicator, and therefore, a corrupted deviation index readout.*

The physicists' formula for the deviation index is

$$DI = 10\log_{10} (EI/EI_T)$$

where EI is the exposure indicator readout for each exposure and *EI_T* is the target EI for "ideal exposure level." Table 10-1 shows that for a range of exposure levels ranging from 1/2 to 2 times the ideal, this formula results in a numerical DI readout scale that extends from −3.0 to +3.0. We've added a column to translate these numbers into *percentages of deviation* from the ideal exposure that will make sense to the radiography student. Note that the ideal range of exposures extends from 20% below the target EI to 25% above the target EI. Why are these two numbers different?

Table 10-1
THE DEVIATION INDEX: ACCEPTABLE PARAMETERS FOR EXPOSURE INDICATORS

Deviation Index	Exposure Deviation	Description	Recommended Action
≥ +3.0	> 100% too high	Excessive patient exposure	**No repeat** unless saturation occurs— Management follow-up
+1 to +3	25% to 100% high	Overexposure	**No repeat** unless saturation occurs
−0.5 to +0.5	−20% to +25%	Target Range	
−1 to −3	20% to 50% too low	Underexposure	Repeat only if radiologist dictates
< −3.0	< 50% too low	Excessive underexposure	**Repeat** (Excessive mottle certain)

Working with percentages can be tricky. As an example, let's change exposure conditions up and down by a factor of 2: Reducing it by a factor of 2 means we cut the exposure in half. Expressed as a percentage, we are at 50% of the original exposure. Now let's increase by a factor of 2 and we double the exposure. Expressed as a percentage this is *not* a 50% increase, but a 100% increase. Everything is clearer when we abandon the percentage expressions and go back to the *factors* or *ratios* by which the exposure was changed, in this example a factor of 2.

If we express the range of ideal exposures as *ratios* rather than percentages, we find that a 25% increase, or 125% of the original is expressed as 5/4 (five-fourths) of the original exposure. A 20% reduction leaves 80% remaining, which is 4/5 (four-fifths) the original. We see that *these are actually proportionate changes.* To reduce exposure by the same *proportion* as a 5/4 increase, we *invert* the ratio and go to 4/5 the original. We can then go back and express these factors as percentages (if we want to make it more complicated). Therefore, *the acceptable range for an "ideal exposure" is from 4/5 to 5/4 of the target EI. In percentages, this is a range from 80% to 125%.*

Now, let's explain how the deviation index formula converts these ratios or percentages into index numbers from −3 to +3. The AAPM explains, "The index changes by +1.0 for each +25% increase in exposure, and by −1.0 for each −20% change." But we do not add one step change to the next, rather, for each added step we *multiply* the result from the previous step (not the original amount) by the factor 1.259 for *increasing* exposure and by 0.794 for *decreasing* exposure. (Note how close these numbers are to the range 80%–125% from the last paragraph.) Table 10-2 shows how these step multiplications work, and the percentage exposure that results: A deviation index of +3.0 indicates a 100% increase to the target exposure, or an increase by a factor of 2, and that a DI of −3.0 indicates 50% exposure or a decrease by a factor of 2.

The right-hand column of Table 10-1 presents the AAPM's recommended actions for each exposure range. An image presenting a DI less than −1 (less than 80% of the target EI) is considered underexposed but *should not be repeated unless a radiologist finds an unacceptable amount of mottle present.* If the DI is less than −3, the exposure to the IR was less than one-half (50%) of the target EI. Assuming that the target EI was properly determined, it has been clinically established that unacceptable levels of mottle can be expected at exposures less than one-half of the EI$_T$, and the exposure should be repeated.

No overexposed digital image should ever be repeated unless saturation *occurs.* When the DI falls within the "overexposed" range of 1.0 − 3.0, *if*

Table 10-2
CALCULATION AND INTERPRETATION OF DEVIATION INDEX VALUE

Deviation Index	Calculated as:	Percentage of Target Exposure
1.0	1.259 = 126%	26% overexposure
2.0	1.259 x 1.259 = 1.585 = 159%	59% overexposure
3.0	1.259 x 1.259 x 1.259 = 1.996 = 200%	100% (2 times) overexposure
−1	0.794 = 79%	79% of target exposure
−2	0.794 x 0.794 = 0.630	63% of target exposure
−3	0.794 x 0.794 x 0.794 = 0.500	50% (one-half of target exposure

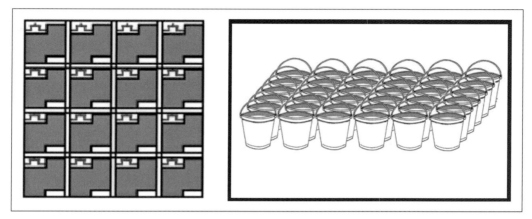

Figure 10-1. *Saturation:* Like filled buckets spilling over into each other, electrical charge can overwhelm the capacity of the dels (left) to store it, spilling across an area of the active matrix array in a DR detector. With all pixel values at a maximum, there ceases to be any distinction between pixels, thus, no image to process.

the image is dark on presentation, the operator should be able to correct this image with windowing controls. If that is still unsatisfactory, alternate algorithms might be applied (see section 1 of this chapter). *Do not mistake a very dark image for a saturated image. If any details at all can be made out in the dark portions of the image, it is over-processed but* not *saturated.*

True saturation is an *electrical* phenomenon that occurs at the *detector* when the dexels in a particular area have reached the maximum electrical charge that they can store. Imagine the DR detector as a "field" of buckets into which water is being poured (Figure 10-1). Each bucket represents a detector element (dexel). When a single bucket is over-filled, it spills water (or electric charge) over into the adjoining buckets and the area becomes saturated. Any further increase in exposure cannot be measured, because all dexels have reached their maximum value. With all values reading out at the maximum, the data becomes meaningless. In the saturated area of a digital radiograph, *all tissues are displayed as pitch black.* As shown in the lung fields in Figure 10-2, saturation represents *a complete loss of data.* These areas of the image are not just "dark" rather, they are pitch black with no details visible.

As digital x-ray machines have been improved, image acquisition has become increas-

ingly robust and true saturation is now a fairly rare occurrence. It has been experimentally demonstrated that even the most sensitive digital units require *at least* 10 times the normal radi-

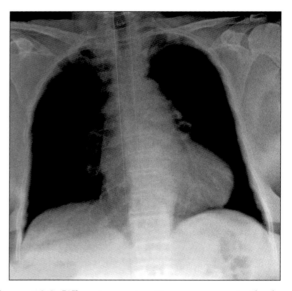

Figure 10-2. When true saturation occurs, as in the lung fields of this chest image, very extreme overexposure results in a complete loss of data, resulting in a flat black area. With simple overexposure, some anatomical details can still be seen, with true saturation, no visible details are present. (From Q. B. Carroll, *Radiography in the Digital Age,* 3rd ed. Springfield, IL: Charles C Thomas, Publisher, Ltd., 2018. Reprinted by permission.)

ographic technique for a particular procedure (assuming average patient size) for saturation to occur. This is more than 3 doublings of the "average" technique. (For some manufacturers, up to 50 times average technique was required to induce saturation!)

Limitations of the DI

Finally, it is important to note that application of the deviation index has its limitations. The AAPM states that it is intended to indicate the acceptability of SNR (signal-to-noise ratio) conditions from *selected values of interest (VOIs) in the image histogram.* Since the EI is derived from the pixel values in the histogram, it is not a *direct measurement of patient dose, and not even a direct measurement of IR dose.* Rather, it is an *indicator* of IR dose, from which patient dose can be roughly extrapolated. (To obtain a scientifically accurate measurement of skin entrance dose (SED) at the surface of the patient, the best approach is to place a solid state (crystalline) radiation monitor or an ion chamber calibrated to diagnostic x-rays at this location and take a direct reading. Such devices are placed within "water phantoms" to estimate internal doses.)

Nor should the DI be taken as the indicator of image quality. Rather, it is one of several indicators. The following quote from AAPM Task Group 116 (*Medical Physics,* Vol. 36, No. 7, July 2009) cautions against inappropriate clinical use of the deviation index:

> Even if images being produced clinically have corresponding DI's well within the target range, the clinical techniques used may still not be appropriate. One can just as readily achieve an acceptable DI for an AP L-spine view with 65 kVp as with 85 kVp; evidence of underpenetration and concomitant excess patient exposure with the lower kVp may be . . . windowed and leveled out in a digital image. Similarly, poor collimation, unusual patient body habitus, the presence of prosthetic devices, or the presence of gonadal shielding in the image may raise or lower DI's (depending on the exam and projection) and perhaps hide an inappropriate technique.

At the very least, any conclusions about image quality must combine the DI readout with visual evaluation of the displayed image and any digital processing features applied. And, any conclusions about patient exposure must combine the DI readout with careful consideration of all pertinent physical factors including patient positioning and the radiographic technique used.

From the previous discussion of the deviation index, one might surmise that a broadly defined range of acceptable parameters for exposure is from one-half to two times the "ideal" exposure amount for each radiographic projection. This is true, and has been used for decades with film radiography, where these were visually "passable" images based on their overall density being neither to light nor too dark for diagnosis, and corresponded to actual exposures at the IR between 5 and 20 microgray, with 10 microgray being ideal. This range of acceptable exposures from 50% to 200% of the ideal correspond to the −3.0 and +3.0 limits for the deviation index.

A Note on "Controlling" Factors

Table 10-3 presents an overview of what might be listed as the "controlling factor" for each constructive image quality. The student will note right away that, from the days of film-based imaging, only *one* of these five factors has *not* been changed by the advent of digital imaging, namely, shape distortion which is still primarily controlled by alignment (positioning) of the x-ray tube, body part, and IR. For film radiography, image density was primarily controlled by the set mAs, contrast by the set kVp, sharpness by the focal spot size, and magnification by the distance ratios between the SID, SOD, and OID. Other factors also affected each image quality, for example, the patient thickness actually had much greater impact on contrast than the set kVp. But, the listed factors were commonly taught as those factors *under the radiographer's control* which were the main determinants for image quality.

It is essential for the radiography student to understand that all of these relationships still

Table 10-3
COMPARISON OF PRIMARY "CONTROLLING FACTORS" FOR DISPLAYED IMAGE QUALITIES

Image Quality	Film-Based Radiography	Digital Radiography
Brightness/Density	mAs	Rescaling
Gray Scale/Contrast	kVp	LUTs (Gradation)
Sharpness (Spatial Resolution)	Focal Spot	Pixel Size
Magnification	Distances (SID/SOD Ratio)	Matrix Size or Field of View
Shape Distortion	Alignment	Alignment

hold true *for the latent image being carried by the remnant x-ray beam to the IR.* The intensity of remnant radiation is still controlled primarily by the set mAs, the contrast of the remnant beam signal by the kVp, the sharpness of the latent image reaching the IR by the focal spot, and magnification by distance ratios. However, once this latent image is passed from the IR into the computer for processing, nearly *everything changes.*

Digital processing is so powerful that processing variables generally have an even greater effect on the final outcome than the conditions of the original exposure. (The initial exposure conditions are still essential to the imaging chain, because they ensure that the latent image being fed into the computer contains sufficient data for digital manipulation, with sufficient discrimination (contrast) between tissue areas for image analysis. But, the fact that these conditions must be met *first* in the chain of events does not mean that they are still predominant at the end of the process when the final image is displayed.)

For the final digital image displayed on a monitor, the average brightness or density level has been primarily determined by *rescaling,* the gray scale and contrast by *LUT's,* the sharpness of the displayed image by the hardware *pixel size* for the display monitor, and the magnification level by the size of the display matrix or the field-of-view (FOV) displayed after adjusting the level of zoom. For this final displayed image, the

initial exposure factors are better thought of as "contributing factors," certainly not as "controlling factors."

Chapter Review Questions

1. An imaging system that requires twice as much radiation exposure than another has _____ the *speed.*
2. For a modern digital x-ray unit, its speed is based primarily upon the digital _____ used rather than the image receptor alone.
3. Because the speed setting for a digital unit can be changed, we refer to it as the system's *speed* _____.
4. Nearly all DR and CR machines can be operated at a speed class of 350-400 without the appearance of substantial _____.
5. Because displayed digital images do not provide any visual cues of overexposure, manufacturers developed various types of _____ _____ to report the level of radiation received at the IR for each exposure taken.
6. For a selected speed class of operation, the "ideal" amount of x-ray exposure at the IR can be designated. This number is called the _____, abbreviated EIt.
7. For most manufacturers, the EIt is derived from the average pixel value or mid-point of the acquired _____.

8. In 2009, to standardize the reporting format of exposure, the American Association of Physicists in Medicine (AAPM) recommended a _____ *index* that should be displayed immediately after each exposure and stored in its DICOM header.

9. Anything that can cause errors in histogram analysis can _____ the accuracy of the DI.

10. The DI readout scale ranges from _____ to _____.

11. In the DI system, the ideal range of exposures extends from _____ percent below the target EI to _____ percent above the target EI.

12. Expressed as *ratios,* this range (in #11) is from _____ to _____ of the target EI.

13. A deviation index readout of −3 indicates _____ of the target exposure was received at the IR.

14. In the "overexposed" DI range of +1 to +3, the operator should be able to correct the image using _____ controls at the console.

15. The only time a repeated exposure is absolutely indicated by a *low* DI readout is when it falls below _____.

16. For DI readouts between −1 and −3, the exposure should not be repeated unless a radiologist finds unacceptable levels of _____ in the image.

17. The *only* time a repeated exposure is indicated for a *high* DI readout is when _____ occurs.

18. True saturation is an _____ phenomenon that occurs at the detector.

19. An image with true saturation is not just dark in appearance, but displays a complete _____ of data in the saturated area.

20. Even the most sensitive modern digital units require at least ___ times the normal exposure to manifest saturation.

21. The EI is *not* a direct measurement of actual patient _____, only an indicator of _____ to the IR.

22. The DI is just *one* of several considerations in evaluating digital image quality.

23. Digital radiograph processing is so powerful that the technique factors used for the initial exposure should now be considered as _____ factors to final image quality rather than as "controlling" factors for the final displayed image.

Chapter 11

DIGITAL IMAGE ACQUISITION

■ ■

Objectives

Upon completion of this chapter, you should be able to:

1. Describe the basic components and function of a DR *dexel (detector element)*.
2. Describe the basic components and function of the *active matrix array* used in DR.
3. Compare the function and advantages between direct-conversion and indirect-conversion DR systems.
4. List the component layers and describe their functions for a CR photostimulable phosphor (PSP) plate.
5. Explain how a CR reader processes the PSP plate to develop a diagnostic digital image.
6. Describe the risks of background and scatter radiation to stored PSP plates.
7. Given a pixel size, calculate the sharpness of a digital image in line-pairs per millimeter (LP/mm).
8. Describe how changes in pixel size, field of view and matrix size do or do not affect image sharpness for the hardware matrix of the display monitor, the hardware active-matrix array of a DR system, the "soft" matrix of the light image from a PSP plate, and the "soft" matrix of the light image displayed on a monitor.
9. State the factors that affect the efficiency of modern image receptors.

10. Describe how the *fill factor* of a dexel affects detective quantum efficiency and sharpness.
11. Explain how *aliasing (Moire)* artifacts can be caused by sampling mismatches in CR readers and on display monitors, and from the use of stationary grids.
12. List the most common artifacts for digital imaging equipment and their causes.

There are two general types of digital x-ray machines in common use for image acquisition in the x-ray room: Digital Radiography (DR) and Computed Radiography (CR). *DR* can be further broken down into *direct-conversion DR* and *indirect-conversion DR.*

The DR unit has the ability to capture, process and display the initial image all within the x-ray unit itself. The image receptor (DR detector) is *directly connected* to the digital processor, so acquired images are automatically, *instantly* sent for processing as soon as they are exposed. This eliminates the need for the radiographer to physically carry an exposed IR cassette from the acquisition unit to a *separate processor*, such as a CR reader, saving time and increasing efficiency.

Mobile DR units may have a cord attached from the main unit to the detector plate, through which the electronic image is transmitted from the IR to the digital processor. These cords are cumbersome and create additional challenges for mobile positioning and "working around the patient." Newer units are cordless. Immediately

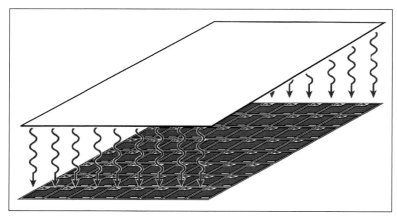

Figure 11-1. For an indirect-conversion DR detector, a phosphor layer (*top*) first converts incident x-rays into light, which then exposes the active matrix array (AMA) of detector elements (dexels) below.

after an exposure, the image is transmitted from the detector plate to the processor using radio waves. We might consider the plate to be "directly connected" to the unit *electromagnetically.* The image is displayed within just a few seconds after exposure, just as with a corded unit.

The main image-capture component of all DR detectors is the *active matrix array (AMA),* a flat panel consisting of thousands of individual electronic *detector elements,* usually referred to as *dexels* or *dels.* The typical size of a single dexel is about 100 microns square. This is one-tenth of a millimeter (about 1/10th the size of a pinhead). This is just at the threshold of human vision at normal reading distance, resulting in an image that appears *analog* at that distance. Because these dexels can be manufactured as extremely thin, flat "chips," the entire AMA can be housed in a panel thin enough to use as a "portable" IR plate for trauma and mobile radiography, called *flat panel technology.*

For *direct-conversion DR* systems, the detection surface of each dexel is made of amorphous selenium, because of its high absorption efficiency for x-rays. The AMA panel converts the energy of the remnant x-ray beam directly into electrical charges which can then be "read out" to the computer. For *indirect-conversion DR* systems, a phosphorescent "screen" is laid over the AMA. This phosphor converts x-rays into light, which then strikes the AMA panel below (Figure 11-1).

The sensitive detection surface of each dexel must now be made of amorphous *silicon,* which is better at absorbing visible light (and not very good at absorbing x-rays). The term *amorphous* ("without shape") means that the selenium or silicon is in a non-crystalline or powder form which allows it to be coated onto each dexel in controlled, extremely thin layers.

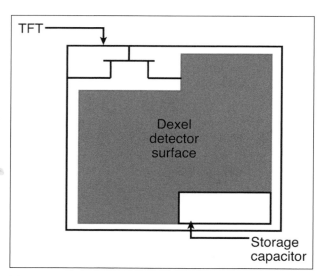

Figure 11-2. Each DR detector element (*dexel* or *del*) is a hardware device consisting of a semiconductor detector surface area, a microscopic capacitor, and a thin-film transistor (TFT) that acts as a switching gate. (From Q. B. Carroll, *Radiography in the Digital Age,* 3rd ed. Springfield, IL: Charles C Thomas, Publisher, Ltd., 2018. Reprinted by permission.)

Direct-Conversion DR Detectors

Every dexel (or *del*) consists of the three basic components illustrated in Figure 11-2. These are 1) a radiation-sensitive detector surface that dominates most of the area of the dexel, 2) a *thin-film transistor (TFT)* "switch," and 3) a small capacitor to store electric charge. The detection surface is a 0.2 millimeter thick layer of semiconductor material that is sensitive to x-rays (for direct conversion) or to light (for indirect conversion). Figure 11-3 diagrams the dexel in cross-section, and we see that for direct conversion, *amorphous selenium* is used to convert x-ray energy into electric charge. This is based upon the simple *ionization* of the selenium atoms by x-rays, where an orbital electron is "knocked out" of the atom by the energy of an x-ray striking it. Each ionizing event creates an *electron-hole pair*, defined as the liberated, free electron and the positively-charged "hole" it has left in the atom, which is now one electron short and therefore positively charged.

The dexel has two charged electrode plates, one above and one below the semiconductor layer. The top electrode is given a positive charge, and the *dexel electrode* at the bottom is charged negatively. By the laws of electrostatics, elec-

trons that are freed from their selenium atoms will tend to drift *upward* toward the top electrode, both pulled by it and repelled by the bottom electrode. Imagine the selenium atoms arranged in perfect *layers,* and we can visualize that when the top layer's electrons move up to the top electrode, positive "holes" are created throughout this layer, which attract any electrons below it. Electrons from the next layer down will be pulled out of their atoms to fill the vacancies in the first layer. Now the second layer is "full of holes," and will pull up electrons from the third layer below. As the electrons from each successive layer are pulled upward to fill holes, holes in the lower layers become apparent. Imagine watching this process like a movie, and we could just as easily see *the gaps or holes in the layers drifting downward* as the electrons moving upward. Electrically, these are *positive charges moving downward toward the dexel electrode,* each positive charge consisting of a "gap" in its atom.

Negative charge is collected at the top electrode and drained off. At the bottom "dexel electrode," positive charge accumulates and moves to the capacitor to be stored (electrons are pulled out of the capacitor, leaving it positively charged). The amount of positive electrical charge built up on the capacitor is propor-

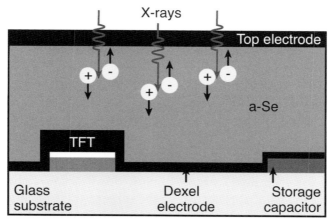

Figure 11-3. Cross-section of a DR dexel during exposure. Each x-ray absorbed by the semiconductor material ionizes a molecule, creating an electron-hole pair. Freed electrons drift upward toward the positive charge of the top electrode, while the "holes" drift downward to deposit a positive charge on the capacitor. (From Q. B. Carroll, *Radiography in the Digital Age,* 3rd ed. Springfield, IL: Charles C Thomas, Publisher, Ltd., 2018. Reprinted by permission.)

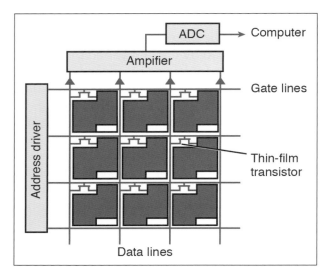

Figure 11-4. Within the active matrix array (AMA), gate lines from the address driver control the sequence for each row of dexels to release their stored electrical charge into the *data lines* leading to the computer. Gate lines apply a small charge to the thin-film transistors (TFTs) to "open the gate" for the stored charge to flow out. (From Q. B. Carroll, *Radiography in the Digital Age,* 3rd ed. Springfield, IL: Charles C Thomas, Publisher, Ltd., 2018. Reprinted by permission.)

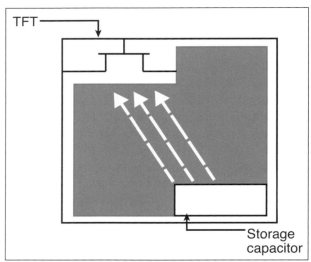

Figure 11-5. When a bias voltage of –5V is changed to +10V in a gate line, a *channel of conductivity* (*arrows*) forms between the capacitor and the TFT, allowing stored electrical charge to flow across and out of the dexel into a data line. (From Q. B. Carroll, *Radiography in the Digital Age,* 3rd ed. Springfield, IL: Charles C Thomas, Publisher, Ltd., 2018. Reprinted by permission.)

tional to the amount of radiation received, a measure of x-ray exposure to the entire dexel.

Figure 11-4 is a diagram of nine dexels embedded within the active matrix array (AMA) of a DR detector plate. Two types of microscopic wires crisscross between the dexels, *gate lines* and *data lines.* The gate lines are connected to an *address driver* which keeps track of the location of each dexel in each row. Each gate line normally has a slight electrical current passing through it with a very low *bias voltage* of –5 volts. To "read out" the information on an exposed DR detector, the bias voltage in the gate line is changed from –5 to +10 volts.

Feeling this sudden change in charge from negative to positive, the TFT for each dexel switches on, effectively "opening the gate" for electricity to flow through. Shown in Figure 11-5, a channel of conductivity through the semiconductor layer opens between the capacitor and the TFT "gate." This allows the charge stored up on the capacitor to flow across the semiconduc-

tor layer, through the TFT gate, and into a *data line.*

The change in the bias voltage of a gate line causes the TFTs to open up sequentially, dumping the stored up charge from each dexel into a data line in succession. As illustrated in Figure 11-4, all of the data lines are connected to an amplifier that boosts the signal, then sends it through an analog-to-digital converter (ADC) into the computer.

Indirect-Conversion DR Detectors

Figure 11-6 illustrates one dexel from an indirect-conversion DR detector. Here, the remnant x-ray beam strikes a phosphor screen (much like the screens used for film radiography) that fluoresces when exposed to x-rays, giving off visible light. The two most common phosphor crystals used are made of cesium iodide or gadolinium oxysulfide. To reduce the lateral dispersion of emitted light, a method was devised of developing the crystals into long, narrow shapes that are vertically oriented, forming *vertical channels* that direct more of the light downward and less sideways. (In spite of this correction, the sharpness of indirect-conversion systems still falls short of that for direct-conversion systems.)

Beneath the phosphor layer, we find precisely the same active matrix array as that used for direct-conversion systems, *except* that the sensitive detection layer for each of the dexels is made of *amorphous silicon* rather than amorphous selenium, because the AMA will now be detecting visible light with a much lower photon energy range than x-rays have. (The orbital electrons of selenium can be dislodged from their atoms by much lower-energy photons.)

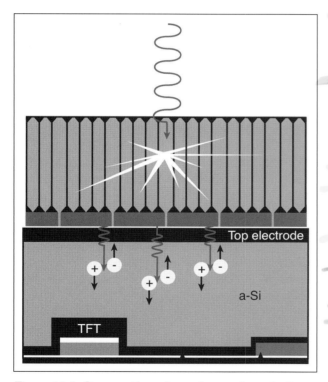

Figure 11-6. Cross-section of one element in an indirect-conversion DR detector. X-rays striking the phosphor layer above are converted into light photons, which then liberate electrons in the amorphous silicon detection surface of the dexel below. (From Q. B. Carroll, *Radiography in the Digital Age,* 3rd ed. Springfield, IL: Charles C Thomas, Publisher, Ltd., 2018. Reprinted by permission.)

Comparing Figure 11-6 to Figure 11-3, we see that the process within the dexel itself is identical to that for a direct-conversion system: Light exposure creates electron/hole pairs. The positively charged holes drift directly downward to the dexel electrode and are conducted to the capacitor for storage. The more electrical charge is built up on the capacitor, the *darker* this pixel will be displayed in the final image. Dexel charges are "read out" by an array of data lines that pass the sequential "bursts" of electric charge from all the dexels through an ADC to the computer.

Computed Radiography (CR)

Computed radiography was developed to simulate film-based radiography, where an exposed "cassette" must be physically carried from the x-ray machine to a processing machine called a *CR reader.* CR cassettes are light and thin for ease of use, especially with mobile procedures, and come in most of the conventional sizes used for film radiography. Like the older film cassettes, they can be placed "tabletop" or in the bucky tray of a conventional x-ray machine. However, these cassettes do not have to be "light-tight" as film cassettes did. The image receptor for CR is a thin *photostimulable phosphor (PSP)* plate with an aluminum backing that is inserted into the CR cassette for the sole purpose of providing rigid structural support while it is placed under a patient or in a bucky tray.

The cassette is made of aluminum or plastic with a low-absorption carbon fiber front panel. A memory chip is placed in one corner to electronically store information on the patient and the procedure. The inside of the cassette is lined with soft felt material to cushion the plate and minimize build-up of electrostatic charges. A small slider button at one end of the cassette releases the PSP plate for removal from a slot. Within the CR reader, this button is moved automatically to retrieve and reinsert the PSP plate. Thus, radiographers directly handle the PSP plate only on rare occasions to investigate a problem or for teaching purposes. *When they do, they must remember that the PSP plate is coated only on one side*

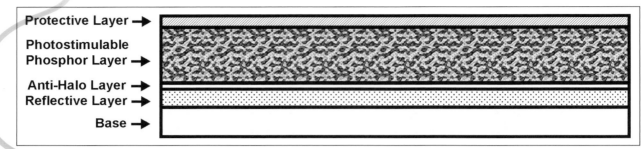

Figure 11-7. Component layers of a photostimulable phosphor (PSP) plate used in a CR cassette.

which must be placed facing forward when it is rein-serted into the cassette.

The basic components of the PSP plate are diagrammed in Figure 11-7. The PSP plate is about one millimeter thick and somewhat flexible. The phosphor layer is supported on a sheet of aluminum, and protected by a thin coat of plastic on top. Between the phosphor layer and the base, there are usually two addition thin layers of material; The lower one is a very thin *reflective layer* of glossy material. Light emitted from the phosphor layer in a *backward direction* is thus reflected back toward the light detectors in the CR reader, improving efficiency. However, a laser light beam is used in the CR reader to stimulate this emission of visible light, and we do not want the laser beam itself to be reflected back toward the detectors. For this purpose, an *anti-halo* layer is placed between the phosphor and the reflective backing. This is chemical which acts as a light *filter,* allowing visible blue phosphor light to reach the reflective layer but blocking the red laser light from penetrating through to it.

Stimulated phosphorescence is a phenomenon observed in certain barium halide (salt) compounds such as barium fluorochloride and barium fluorobromide. Activated with small amounts of europium, an originally pure crystal of these compounds will develop tiny defects called "metastable sites" or "F centers" (from the German for *Farbzentren* ("color centers"). Each F center has the ability to "trap" free electrons and store them until some form of re-stimulation dumps additional energy into the F center to "excite" it. During an x-ray exposure, atoms within the barium-halide molecules are ionized, and some of the freed electrons become trapped in the F centers (Figure 11-8).

(At the same time, the phosphor screen also fluoresces, giving off a beautiful blue-violet light that can be demonstrated by direct observation. This light represents the majority of freed electrons, which immediately "fall back" into their atomic orbits, emitting light waves in the process. However, a small percentage of freed electrons are trapped in the F centers for processing at a later time.)

Placed in a CR reader (processor) the PSP plate is removed from its cassette and scanned by a helium-neon red laser beam that moves across the plate and then indexes down one row at a time. At each F center, electromagnetic energy from the laser beam is added to trapped electrons, speeding them up in their vibrations and enabling them to "jump" out of the trap. Thus freed from the F centers, these electrons now fall back into the shells of local atoms that have vacancies. As they settle into these atomic orbits, the potential energy they lose is emitted as light. This is the *second time* the phosphor plate has glowed, and because it consists of such a feeble number of remaining electrons, the light is quite dim. This image must therefore be electronically amplified before it can be displayed as a radiograph.

The immediate emission of light by a substance under some type of stimulation is called *fluorescence,* but some substances can store the energy from stimulation for a period of time before emitting it back out in the form of light or "glowing in the dark." The proper term for

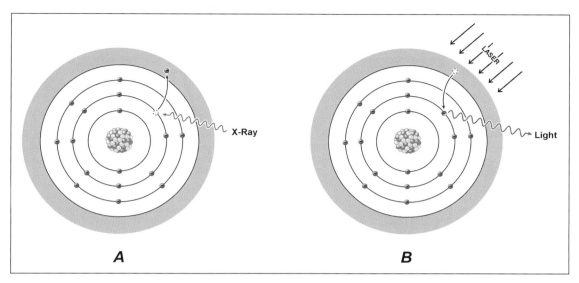

Figure 11-8. X-ray exposure ionizes orbital electrons, raising them to a higher energy level, **A**. In the photostimulable phosphor (PSP) plate used for CR, some of these liberated electrons become trapped in an energy band called the *F Center (gray area),* an energy band created by the peculiar molecular structure of the PSP. When the PSP plate is processed in a CR reader, a laser beam is used to free trapped electrons from the F centers, **B**, and as they fall back into their atomic orbits, energy is emitted in the form of light.

this delayed emission of light is *phosphorescence.* With CR, x-ray exposure constitutes a "first stimulation" of the phosphor plate resulting in immediate fluorescence, and then a second stimulation is applied by the reader's laser beam after a time delay. Hence, the term *stimulated phosphorescence.* This is phosphorescence or delayed light emission, but stimulation by laser was required to induce it.

The CR Reader (Processor)

The CR reader uses suction cups and rollers to mechanically retrieve the PSP plate from the CR cassette and move it from one station to another. In the reading chamber, a laser beam is deflected off a rotating mirror, as shown in Figure 11-9, to make it scan across the PSP plate from side-to-side. When this row is completed, rollers advance the plate by a tiny amount and the laser makes another sweep. The crosswise direction in which the laser beam moves is referred to as the *fast scan direction.* The direction in which the PSP plate itself is advanced by the rollers is called the *slow scan* or *subscan direction.*

The laser beam is in the shape of a circle with an 80-micron diameter (just under 1/10th of a millimeter or 1/10th the size of a pinhead). Due to its circular shape, the laser spot must overlap pixels that will be displayed roughly as squares on a monitor. This situation is worsened at the far ends of each row where the angle of the laser beam distorts it into an oval rather than a circular shape.

DR has the advantage of using square hardware dexels with well-defined borders, however, it presents the disadvantage of missing some detection area in the form of spaces between dexels and the space taken up by capacitors and TFTs, reducing efficiency.

Unlike DR, the location and size of pixels in CR is defined *during processing* rather than by the image receptor. The PSP plate in the CR cassette has no defined pixels—in fact, it is an *analog* image in every way. The *maximum* size of the pixels is defined in the CR reader by the width of the laser beam. Their *location* is defined by the *sampling rate,* that is, how many individual measurements are taken across each row or scan line. We can only increase the number of sam-

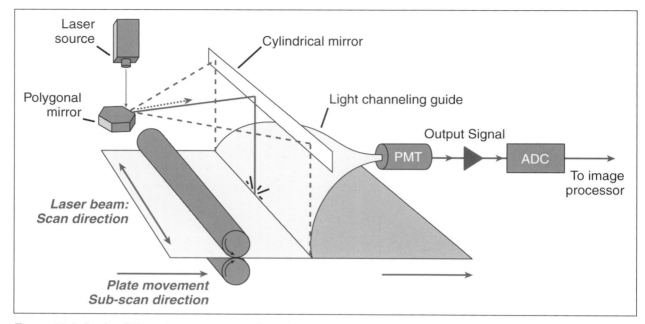

Figure 11-9. In the CR reader, a rotating polygonal mirror is used to sweep the laser beam across the PSP plate. The "fast scan" direction is the crosswise laser beam scan direction. The "slow scan" or "subscan" direction is the direction of the plate as it moves through the reader.

ples taken across a particular line by *reducing the distance between samples.* Taking samples more frequently forces us to take them *closer together.* This can only be done by defining them as smaller pixels insofar as the final image is concerned. The sampling rate determines the location, and ultimately, the size, of CR pixels—the *fast scan* sampling rate sets their width, and the *slow scan* sampling rate sets their length. (To keep pixels from being distorted into ovals rather than circles, the fast scan and slow scan sampling frequencies must be equal.)

At each predetermined pixel location, as the laser beam stimulates the phosphor crystals in the area, a discrete measurement is taken of the quantity of visible light emitted by the pixel. To do this, the light is first directed through a *light channeling guide* to a *photomultiplier tube* (Figure 11-9). The light intensity is so dim that it must be amplified to produce a readable signal. In the photomultiplier (PM) tube (Figure 11-10), light energy is first converted into a stream of electrons by a *photocathode* plate. All photocathodes use chemical compounds that are highly

susceptible to being ionized by light photons, predominantly by the photoelectric effect.

In the PM tube, the intensity of the stream of electrons is multiplied over and over again by passing it through a series of *dynodes,* electrodes that can be quickly switched back and forth between positive and negative charge. Just as the electrons reach a dynode, it switches from positive to negative, thus repelling the stream toward the next dynode. The acceleration energy gained by the stream liberates an additional 5 electrons from each dynode plate for every electron striking it, until a greatly multiplied number of electrons reaches the collecting plate. This results in one strong pulse of electricity that can be measured and manipulated.

The PM tubes in a CR reader are specifically sensitive to blue-violet light wavelengths, the color emitted by the PSP plate. This prevents any reflected *red* light from the laser beam from interfering with the emitted light signal.

After scanning is completed, the PSP plate is moved by rollers into the *eraser* section of the reader. Here, the plate is exposed to intense

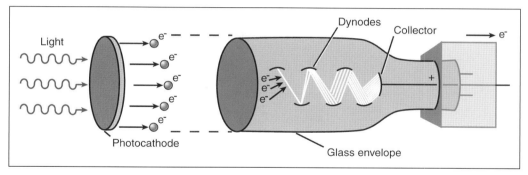

Figure 11-10. During CR processing, light emitted by the PSP plate is directed by a light channeling guide (see Figure 11-9) to a photomultiplier (PM) tube. Shown here, the PM tube first uses a photocathode layer to convert light energy into an emitted stream of electrons (by ionization), then the electron stream is greatly amplified by a series of dynodes that sequentially boost the signal. (From Q. B. Carroll, *Radiography in the Digital Age*, 3rd ed. Springfield, IL: Charles C Thomas, Publisher, Ltd., 2018. Reprinted by permission.)

white light to remove any remaining information from the plate so it can be reused. The plate is then reloaded into its cassette and ejected from the machine. The entire reading process takes about 90 seconds to complete.

Every CR cassette includes a "blocker" area with electronic memory, used to imprint the date and identification information on the patient, the institution, and the procedure prior to exposure. This information is entered by computer keyboard and then "flashed" on the cassette electromagnetically. Each cassette also has a unique identifying number for tracking, that can be accessed the same way or, more quickly, by a bar code using a bar code scanner.

Background and Scatter Radiation

The PSP plates used for CR are very sensitive to background radiation. It only takes 1 milligray exposure to create a noticeable "fog" density on a CR plate, and background exposure can be up to 0.8 milligray *per day.* Therefore, CR plates should always be erased in the CR reader prior to use if there is any chance they have been in storage for more than a couple of days, such as over a weekend. If a CR cassette is left against a wall in an x-ray room, even at a considerable distance from the x-ray machine, it can quickly accumulate enough scatter radiation exposure from repeated use of the room to fog the plate. CR cassettes must be stored in lead-shielded areas such as behind the control booth, and must be erased prior to use if there is any question of exposure to scatter radiation.

Such "pre-fogging" events ensure that the lightest densities in the image *histogram* will be gray rather than "white," and reduce the contrast-noise ratio (CNR) for the latent image. If enough contributing factors generate sufficient noise to impair histogram analysis, segmentation or processing errors can result in a light or dark image being displayed. (These types of "pre-fogging" events are not to be confused with scatter radiation created *during an exposure* to a "clean" PSP plate. In Chapter 4, we demonstrated that digital processing is generally robust enough to correct for the effects of scatter produced *during the exposure,* almost to a miraculous degree.)

For DR detectors, after each use, a "flash" electronic exposure is used to purge the detector plate of all residual charges remaining within the dexels. Since the detector is effectively *erased* before each x-ray exposure, the problems of background and scatter radiation are obviated.

Sharpness of Digital Systems

We have mentioned that in a CR reader, the only way to increase the sampling frequency,

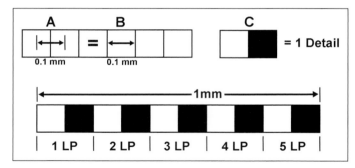

Figure 11-11. Pixel pitch or dexel pitch, **A**, is approximately equal to pixel or dexel width, **B**. We generically define an image "detail" as a pair of pixels with different values, **C**. At 0.1 mm in width, the spatial frequency will be 5 line-pairs per millimeter (bottom). This is 5 details per millimeter for each row of pixels.

the number of samples taken per line, is to reduce the distance between them. The distance from the *center* of one pixel to the *center* of an adjacent pixel is called the *pixel pitch*. (For a DR detector, the distance from the center of one hardware dexel to the center of the next one is the *dexel pitch*.) In Figure 11-11 we see that generally, pixel pitch is equal to pixel width or the *size* of the pixels. Both are related to the displayed sharpness of the image.

If we ask, "What is an image detail," in its simplest form we could define a single detail as an "edge" between two pixels with different values, such that one is displayed lighter and the other darker. (If two adjacent pixels were both white, no information or detail would be present.) The upshot of this is that *it takes a minimum of two pixels (with different values) to make a detail.* In Figure 11-11 we see that a row of 10 pixels can display or resolve a maximum of 5 details, each detail consisting of a pair of pixels, one white and one black (or gray). So, the number of resolvable details is *one-half* the number of available pixels. And, the number of available pixels depends upon their *size*.

To measure the spatial resolution of an image or imaging device, we use the unit of *spatial frequency* which is *line-pairs per millimeter*, abbreviated *LP/mm*. The entire row of pixels in Figure 11-11 fits with one millimeter of length. This is 10 pixels or 5 *pixel pairs* per millimeter. Imagine the alternating black and white pixels as *vertical*

lines, and we surmise that the spatial frequency illustrated in Figure 11-9 is *5 LP/mm.*

We can now state the *formula* for spatial frequency. It is

$$SF = \frac{1}{2P}$$

Where "P" is pixel size. Using Figure 11-11 to illustrate, since there are 10 pixels per millimeter, the pixel size must be 0.1 mm. Plugging this number into the formula as "P," we obtain

$$SF = 1/2 \times 0.1 = 1/0.2 = 5 \ LP/mm$$

If the pixels are 0.1 mm wide, the spatial frequency will be 5 line-pairs per millimeter. An image with SF = 10 LP/mm will appear *twice as sharp* as an image with SF = 5 LP/mm.

For a DR detector, the same formula can be applied inserting the *dexel size* as "P." A typical dexel is 100 microns in width—this is 1/10th mm and fits the diagram in Figure 11-11, having a maximum resolution of about 5 LP/mm. Because of other complicating factors, such as the spaces between dexels for DR and light diffusion in CR, the actual resolution that is clinically measured is slightly less than calculations from this formula would indicate, but it gives us a good approximation of what to expect from different imaging systems.

Hardware Matrices

The size of the matrix and the field of view *can* affect spatial resolution or sharpness in the image, *if they alter the size of the pixels or dexels.* Here, we must make a distinction between the *hardware* matrices (matrixes) that make up a DR detector or a display monitor screen, and the soft or variable matrices that make up a displayed light image or the light image emitted from a PSP plate within a CR reader during laser stimulation:

For any *hardware matrix,* the size of the dexels or pixels is *fixed* and not subject to change. For example, the dexels of a typical DR detector are 100 microns in size. This sets the DR detector's *inherent resolution* at about 5 LP/mm by the above formula, and it is not subject to change because it is based on the way that the manufacturer constructed the DR detector, not on the matrix size nor the field of view (FOV). If a physically larger DR detector plate were used, (41 x 41 cm instead of 35 x 35 cm), we could argue that the matrix size has been increased, as shown in Figure 11-12. But, the size of the hardware dexels themselves remains the same. In

this case, *sharpness* has not been altered by using a "different matrix size."

The same holds true for the *hardware pixels of a display monitor screen.* As shown in Figure 12-8 in the next chapter, each "hardware pixel" consists of the intersection of two wires in the form of flat, transparent conductors. Thousands of these flat wires crisscross throughout the monitor screen to make up the display surface. The *size* of each pixel can be measured as the square created by one flat wire crossing over the other. It is *inherent* to the construction of the monitor and is not subject to change, and it sets the *maximum resolution* with which any image can be displayed on that monitor (again, by the above formula). For a particular manufacturer, a physically *larger* monitor (50 cm wide rather than 40 cm wide) will have more hardware pixels in its matrix, so it could be argued that the matrix size has increased. But, *the size of each hardware pixel has not changed,* and spatial resolution is consistent from one monitor to another for this manufacturer.

Now, let's take up the question of *field of view (FOV)* in these two contexts, DR detectors and display monitors. For a DR detector, *collimation*

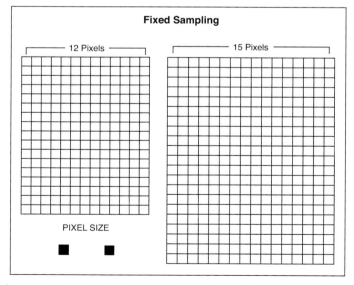

Figure 11-12. In any *hardware* matrix, such as a DR detector plate, dexel size remains the same regardless of the size of the matrix, field of view, or image receptor. Therefore, the inherent sharpness (or spatial resolution) of the device remains unchanged. This is also true for the *hardware* pixels of a display monitor.

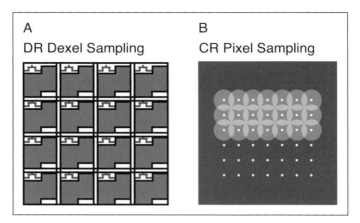

Figure 11-13. For a DR detector, (*left*), dexels are well defined and their size is fixed, but for a CR image (*right*), pixels are round and therefore must overlap each other, and pixel size can be changed by effectively "collimating" the area struck by the laser beam. (From Q. B. Carroll, *Radiography in the Digital Age,* 3rd ed. Springfield, IL: Charles C Thomas, Publisher, Ltd., 2018. Reprinted by permission.)

of the x-ray beam results in a smaller physical area being recorded on the DR detector plate. When this collimated image is displayed on a monitor, we observe a magnified but more restricted field of view. Less of the patient's body is displayed. The effects on the displayed image will be discussed shortly, but here, we are focused on the *DR detector itself.* The *inherent* resolution power of the DR detector is based solely on the size of its hardware dexels. This is not changed by simply collimating the x-ray beam. It is independent of changes to the FOV from collimation.

At the display monitor, if this collimated image is then displayed to *fill the monitor screen,* it would have to first be subjected to some degree of *magnification,* to be discussed shortly. This magnification level of the *light image* can also be changed at the whim of the operator by using the "zoom" feature. But, none of this changes the physical size of the *hardware pixels* in the monitor, so it has nothing to do with the *inherent resolution* of the monitor itself, which is always consistent.

To summarize, *hardware elements in DR detectors and display monitors have a fixed size, so these devices have consistent inherent spatial resolution, regardless of changes to matrix size or FOV.*

The **Soft** *Matrix of a Light Image*

When we examine the *light image* itself which is coming from a display monitor or from a PSP plate under stimulation within a CR reader, we find that the matrix size of the image is *soft,* that is, it can be changed within the physical area of the display. (Shown in Figure 11-13, the CR pixel is actually a round "spot" whose diameter is determined by the width of the round laser beam striking the PSP plate, and can be effectively "collimated" to change pixel size.)

When the matrix is variable *within a given physical area,* it will have an effect on both field-of-view and on *pixel size,* which is now also *"soft"* or variable. Let's begin with the relationship between the matrix and pixel size. Figure 11-14 shows two different matrix sizes that are both forced into a given physical area 3 cm wide by 4 cm long. The only way to fit more pixels within this area is to make them smaller. Therefore, *for a given physical area,* the larger the matrix, the smaller the pixels.

Since smaller pixels result in a sharper image, we conclude that *for a displayed light image, the larger the matrix size, the higher the sharpness.*

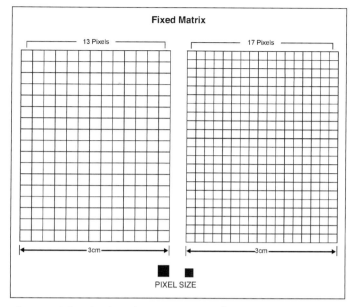

Figure 11-14. If the physical area of the displayed image is fixed (here at 3 cm wide by 4 cm tall), the only way to fit a larger matrix with more pixels into the physical area is to make the pixels smaller. This results in improved sharpness (or spatial resolution).

On a display monitor, the displayed light image can be magnified by the *zoom* feature. As illustrated in Figure 11-15, the first step of magnification consists of *spreading out* the pixel value for each single image pixel across *four hardware pixels* on the monitor. The next step would spread that value across *nine* monitor pixels, and the next step would distribute it over *sixteen* monitor pixels. The hardware pixels of the monitor do not change, but the *visually apparent pixels of the light image itself are each getting larger.* With excessive magnification, we begin to see a *pixelly* image, in which the individual pixels themselves become apparent to the human eye and the image takes on a distractingly "blocky" and blurred appearance.

For a given physical screen size, the more magnification is applied, the smaller the field of view displayed, since less anatomy "fits" within the screen. We conclude that *for a displayed light image, the smaller the field of view, the poorer the sharpness* as the visually apparent pixels become enlarged.

We have learned that within a CR reader, the *pixel size* can be varied by changing the sam-

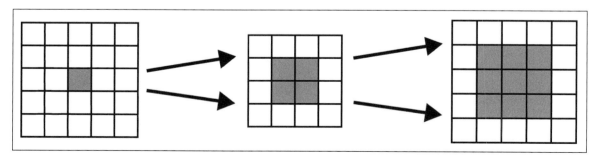

Figure 11-15. To magnify a displayed *light image,* the pixel value for a single pixel is spread out across four *hardware pixels* of the display monitor. Additional magnification spreads the value out over nine *hardware pixels.*

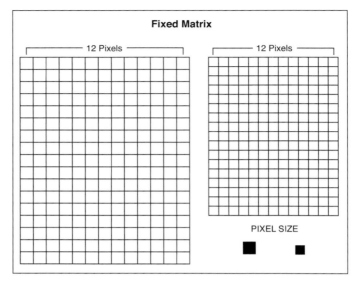

Figure 11-16. In a fixed matrix system, each row is divided into the same number of pixels regardless of the physical size of the image or detector. One result is that smaller CR cassettes produce smaller pixels with increased sharpness (spatial resolution).

pling rate. Here, we start again with a fixed physical area. Within that area, if the sampling frequency is increased so that more pixels are sampled from each scanned line, those pixels must become smaller. Since the physical area is fixed, making the pixels smaller means that more pixels "fit" per line, making the *matrix larger.* Pixel size, matrix size, and FOV can all be variable in CR.

CR units are divided into two broad categories: *Fixed matrix* systems and *fixed sampling* systems. A fixed matrix CR system follows the layout in Figure 11-16. Each row of the image is divided into the same number of pixels per row regardless of physical size of the image. Smaller CR cassettes have smaller PSP plates in them. A smaller PSP plate displays a physically smaller image inside the CR reader. Since each row of this image is sampled the same number of times as for a larger image, but the rows are shorter, the pixels must be smaller to get the same number of samples. The result is that smaller CR cassettes produce sharper output images than larger cassettes.

Nearly all CR manufacturers now produce *fixed sampling* systems that follow the format in Figure 11-12. Regardless of the size of cassette used, the sampling rate is always the same, resulting in consistent pixel size and consistent sharpness.

The Pixel Size Formula

For *light images* being displayed on a monitor or being emitted from a PSP plate in a CR reader, soft matrices apply and pixel sizes can vary within the same physical area. Under these circumstances, a *formula* can be written relating the pixel size to both the matrix size and the FOV. It is:

$$\text{For a given physical area: PS} = \frac{\text{FOV}}{\text{Matrix}}$$

where "PS" is the pixel size and "FOV" is the field of view. For example, what is the pixel size for a PSP plate measuring 205 mm across, divided into a matrix of 1024 x 1024 cells?

$$\textit{Answer: } \text{PS} = \frac{205}{1024} = 0.2 \text{ mm pixel size}$$

Efficiency of Image Receptors

It is the job of both CR phosphor plates and indirect-conversion DR phosphor plates to convert the energy from x-rays into *light.* An "average" x-ray photon in the beam can have 30 kilovolts of energy, while visible light photons have an energy of about 3 volts. This means that *each* x-ray could potentially be "split" into 10,000 photons of light by a phosphor molecule. When we break down this process, we see three critical steps: First, the phosphor layer must be efficient at *absorbing x-rays.* This can be improved by using chemical compounds with a very high *average* atomic number, so that many photoelectric interactions occur, or by using thicker layers of phosphor which has the negative side-effect of reducing sharpness. (The front panels of CR cassettes and DR detector plates must be made of *low x-ray absorption* materials so that x-rays get through to the phosphor or detector elements.)

Next, the phosphor layer must be efficient at *converting x-rays into light.* This is purely a function of the particular chemical compound used. Third, the phosphor plate must be efficient at *emitting the light.* The light must "escape" through the crystals above it and the protective coating. The reflective backing in a CR cassette reverses light emitted in a backward direction and directs it upward, improving emission efficiency.

For DR detectors, recall that the individual detector elements (dexels) must absorb and measure x-rays, yet their sensitive surfaces are very thin. Only a small percentage of the x-ray beam is absorbed by these surfaces, so they must use chemical compounds that are as efficient as possible for absorption.

An indirect-conversion DR system first converts the x-rays into light using a phosphor layer. Each x-ray can yield up to 10,000 light photons, but only a few percent of these are directed downward to the AMA. Still, this would amount to 300 light photons for every x-ray. Furthermore, since these light photons only possess a few volts of energy each, they are not very "penetrating," but are easily absorbed by the amorphous silicon in the dexel. It is for just this reason that indirect-conversion systems require less radi-

ographic technique and save patient exposure. For direct-conversion systems, higher patient dose is a disadvantage, but this is offset by higher sharpness because there is no light dispersion (spreading) to contend with. Both systems continue in clinical use.

Once x-ray or light photons are absorbed in a dexel, both the *conversion* of this energy into electric charge, and the "emission" of this *electronic signal* are nearly 100 percent.

Detective Quantum Efficiency

Detective quantum efficiency (DQE) is used by physicists to measure the overall efficiency with which an imaging system can convert input x-ray exposure into useful output signal (a useful image). Mathematically it is simply the input signal-to-noise ratio (SNR) *squared,* divided by the output SNR *squared.* No imaging system can achieve a perfect DQE of 1.0 (100%). For a typical extremity procedure, the DQE for CR is about 30 percent, for direct-conversion DR about 67 percent, and for indirect-conversion DR about 77 percent. DQE is but *one* measure of an imaging system's overall quality. System latitude, display quality, and several other factors affect both patient dose and image quality.

Fill Factor

For the dexels in a DR detector, an important aspect called the *fill factor* affects the detective quantum efficiency. The fill factor is defined as the percentage of the square area of the dexel that is devoted to the semiconductor detection layer of amorphous selenium or amorphous silicon. A high fill factor provides more contrast resolution, better signal-to-noise ratio (SNR), and improved sharpness. This relationship limits the degree to which we can miniaturize dexels, because the TFT and the capacitor always require the same amount of space and cannot be shrunk in size. Therefore, as shown in Figure 11-17, a smaller dexel will have a reduced fill factor, that is, less detector surface area, and this can require an increase in radiographic technique resulting in higher patient exposure.

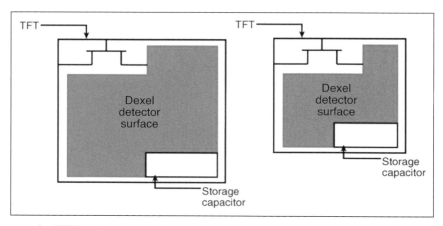

Figure 11-17. Because the TFT and capacitor cannot be reduced in size, smaller dexels have less detection surface area, or a lower *fill* factor. This results in lower spatial resolution, lower contrast resolution and lower sensitivity, which requires increased technique. (From Q. B. Carroll, *Radiography in the Digital Age,* 3rd ed. Springfield, IL: Charles C Thomas, Publisher, Ltd., 2018. Reprinted by permission.)

Digital Sampling and Aliasing

We've discussed the importance of sampling frequency for sharpness. For any particular image, certain sampling frequencies can also cause geometric *artifacts* in the image known as *Moire patterns* or *aliasing*. A Moire pattern is a series of false lines created in the image that can take on many shapes and configurations. They can be curved wave-like patterns as in Figure 9-11 in Chapter 9, or a sequence of straight lines as in Figure 11-18.

The *core* cause of aliasing artifacts is *insufficient sampling of high frequency signals*. Figure 11-19 illustrates this concept. Here, the signal we are trying to detect is a series of alternating black and white pixels. As we learned in Chapter 8, such an image can be represented as a series of waves with alternating *up* and *down* pulses (up for black, down for white). If a sample (mea-

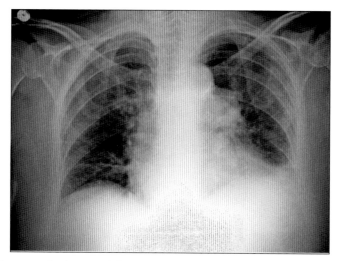

Figure 11-18. Aliasing (Moire artifact) caused by insufficient CR reader sampling of a high-frequency acquired image, resulting in false line patterns. (From Q. B. Carroll, *Radiography in the Digital Age,* 3rd ed. Springfield, IL: Charles C Thomas, Publisher, Ltd., 2018. Reprinted by permission.)

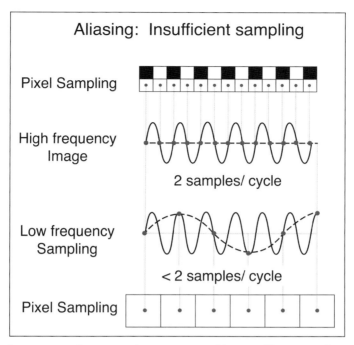

Figure 11-19. In the digital age, most aliasing artifacts are caused by *insufficient sampling* of a high-frequency image (*top*). When copied at 2 samples per cycle (*upper waveform*), the image is properly resolved. If it is copied at *less than 2 samples per cycle* (*bottom waveform*), aliasing lines will appear at each indicated node along the dashed wave line. (From Q. B. Carroll, *Radiography in the Digital Age*, 3rd ed. Springfield, IL: Charles C Thomas, Publisher, Ltd., 2018. Reprinted by permission.)

surement) is taken at *every* pulse (Figure 11-19, *top*), the resulting copy of this image will be clear and artifact-free. If *more* samples than necessary are taken, such as two samplings per original pixel, the sampling detector has higher frequency than the original image and there is no problem reproducing the image, with high fidelity.

However, the bottom of Figure 11-19 shows that if the sampling rate is *less* than the "pulse rate" of the original image, the measurement sampling points become *misaligned* with the image pixels. A false line will be artificially created at each point where the longer-wavelength sampling frequency *overlaps* the shorter-wavelength image frequency. This is the Moire artifact.

There are many variations on how "insufficient sampling" can occur: *Any time a copy of a digital image is made*, aliasing can occur. Examples include not only "copying machines" of various sorts, but more commonly, taking a photograph with a camera or cell phone of a digitally displayed image on another device such as a computer display monitor. A familiar form of "dynamic" or moving moire patterns is caused when the spokes of a rotating wheel reach a certain speed and appear to be moving slowly *backward*. All these effects are due to overlapping sampling rates.

In the computer age, aliasing is also a familiar and common occurrence when images on a digital display monitor are repeatedly "zoomed in" or magnified. At some point, when the sampling rate to create the magnified image becomes less than 2 times the frequency of hardware pixels in the display screen itself, moire artifacts will appear.

The *Nyquist frequency* is the minimum required sampling rate of any image detection device, such as the laser beam scanning across a PSP plate in a CR reader. The *Nyquist theorem* or

Nyquist criterion states that to prevent aliasing, the sampling rate (Nyquist frequency) must be *at least* double the spatial frequency of the image. This is just another way of saying that *each pixel of the image must be sampled.* Remember that in a continuous waveform, it takes *2 pulses* (one up, one down) to make one whole *cycle.* The unit for frequency, the *hertz (Hz),* is defined in whole cycles (e.g., cycles per second), each cycle consisting of two pulses. So, when we say "double the frequency," we're simply saying "equal to the number of pulses (or pixels)."

Aliasing from Grids

For CR, when a stationary grid is used to make the original exposure, the grid can act as a "high-frequency signal" since, in effect, it is "screening" the image between thin lead plates (grid lines). The *grid frequency* (the number of grid lines per cm) interacts with the *Nyquist frequency* of the CR reader. If the grid frequency is less than the CR sampling frequency, aliasing can occur, as shown in Figure 11-20. To prevent this type of aliasing, there are at least three options: 1) conventional grids with very high frequencies can be installed, but these are quite expensive; 2) special "hole" grids can be purchas-

es that use holes rather than slits to screen the x-ray beam, but these are also expensive; 3) non-grid approaches can be used in many cases for the initial exposure, or *virtual grid software* can be installed which obviates the need for grids in most procedures (fully discussed in Chapter 4).

Other Digital Artifacts

In Chapter 5 we discussed dexel drop-out as a form of noise in the image. The resulting "electronic" mottle is much more common for DR than for CR systems. For CR, the most common source of various artifacts is the PSP plate and cassette. These consist of scratches, dirt or dust on the PSP plate, and *ghost images* from a previous exposure remaining on a plate that was not fully erased. Artifacts that run across the entire length of the image can be caused by foreign objects on the laser-deflecting mirror or light guide tubes in the reader. Drop-out of whole lines can be caused by malfunctions (jamming) in the scanning and PSP plate transport systems in the CR reader.

Remember that artifacts appearing on many images taken with various CR cassettes indicate problems inside the CR reader. For both DR and CR, default processing artifacts can occur

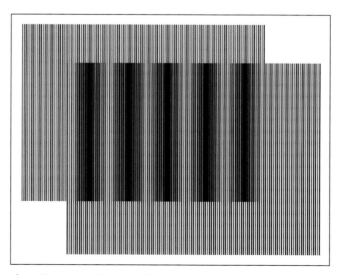

Figure 11-20. Aliasing (moire) artifact caused by grid lines overlapping sampling lines in a CR reader when the grid frequency is too low.

from segmentation failure, corruption of the histogram, improper application of special processing features (such as edge enhancement), and other software problems discussed in Chapters 7–9. *Exposure* artifacts include a wide variety of extraneous objects that might be found in or on the patient or IR during the initial radiographic exposure. Due care must be taken during positioning of the patient to prevent these.

Chapter Review Questions

1. The main image-capture component of all DR detectors is the *active* _____ _____

2. A typical dexel (del) is a square of about _____ _____ in size.

3. For an indirect-conversion DR system, the detection layer of the dexel is made of amorphous *silicon* rather than *selenium* because silicon is better at absorbing _____.

4. List the three main components of a dexel (del):

5. In a dexel, when "electron-hole pairs" are formed from x-ray ionization, the positively-charged _____ drift downward to the dexel electrode, increasing positive charge on the capacitor.

6. List the two types of microscopic wires that crisscross between the dexels of the AMA.

7. When the bias voltage applied to a *thin-film transistor (TFT)* is changed to a positive voltage, the TFT acts as a "_____," opening to allow charge stored on the capacitor to surge out of the dexel.

8. Even though the phosphor crystals of an indirect-conversion DR system are shaped into vertical channels to minimize light dispersion, the sharpness still falls short of that for a _____-_____ DR system.

9. The more electrical charge is built up on a dexel's capacitor, the _____ the corresponding pixel will be displayed.

10. What does *PSP* stand for?

11. When a PSP plate is removed from a CR cassette, it must be reinserted facing _____.

12. The anti-halo layer of a PSP plate reduces reflection of the red _____ light toward the light channeling guide.

13. Even though the PSP plate glows during x-ray exposure, later stimulation by a laser beam frees up residual electrons trapped in molecular _____-centers, causing the plate to glow a second time.

14. The phosphorescent glow of a PSP plate occurs because as freed electrons fall back into atomic shells, they _____ potential energy.

15. Rollers in a CR reader move the PSP plate itself in the _____-scan direction.

16. In a CR reader, the *maximum* size of the pixels produced is determined by the width of the _____ _____.

17. In a CR reader, because the stimulated glow of the PSP plate is so dim, this light is channeled to _____ tubes for amplification.

18. CR plates that have been in storage for more than a couple of days should be erased prior to use because of their sensitivity to _____ _____.

19. Define *pixel pitch:*

20. In the formula for spatial frequency, the number of pixels must be doubled because it takes at least _____ pixels to make up an image detail or one whole cycle of the frequency.

21. The *inherent* sharpness of a display monitor is *not* affected by changes in field of view or matrix size, because the _____ of the monitors' *hardware pixels* is fixed and cannot be changed.

22. The above also holds true for a DR detector and its dexels, which have an inherent resolution of about _____ LP/mm.

23. On a display monitor, increasing magnification of the displayed light image _____ each single pixel value out across several hardware pixels.

24. For a light image displayed on a given size of monitor screen, the larger the matrix size of the light image itself, the _____ the sharpness.

25. As one continues to magnify a displayed image more and more, name *two* problems with its appearance that will inevitably occur at some level of magnification:

26. Which type of CR system is now most common and produces consistent image sharpness regardless of the size of the IR used?

27. The reflective layer in a PSP plate improves which of the three types of IR efficiency?

28. If the dexels in a DR detector are made too small, DQE is lost and radiographic technique must be increased, because the smaller dexels have a reduced _____ factor for the sensitive detection area.

29. The *core* cause of all aliasing (moire) artifacts is insufficient _____ rate of high-frequency signals.

30. The _____ *criterion* states that to prevent aliasing artifact, the sampling frequency must be at least _____ the frequency of the image.

Chapter 12

DISPLAYING THE DIGITAL IMAGE

■ ■

Objectives

Upon completion of this chapter, you should be able to:

1. Distinguish between *diagnostic workstations, workstations* and *display stations.*
2. Describe the *hardware pixel* used in a liquid-crystal display (LCD) monitor and how it uses light polarization and electrical charge to display different levels of brightness (density).
3. List the LCD components needed to produce and properly diffuse light for the displayed image.
4. Define *active-matrix LCD (AMLCD), input lag,* and *dead and stuck pixels.*
5. State the major advantages and disadvantages of liquid-crystal display
6. Describe the sharpness and luminance qualities of the displayed LCD image and how they are monitored.
7. Describe the requirements for ambient lighting in a diagnostic reading room.

Nearly all radiographic images are now viewed on *liquid crystal display (LCD)* monitors attached to a computer that gives the operator the capability of windowing or otherwise manipulating the image as it is viewed. A *workstation* is a computer terminal that can not only access image for display, but can be used to adjust image quality and permanently *save* those

changes into the PAC (Picture Archiving and Communication) system. Patient information attached to each image can also be added or deleted.

Within a radiologist's reading room (Figure 12-1), the *diagnostic workstation* typically consists of *three* display monitors: Two of these are large high-resolution (2000-pixel) monochrome (black-and-white) display screens for viewing radiographs. Two monitors are necessary to allow the essential diagnostic function of comparing two PA chest images, for example, taken at different times, such as pre-surgery and post-surgery, in order to track a disease or condition. This also allows the radiologist to examine a PA chest and lateral chest on the same patient side-by-side as shown in Figure 12-1. The third monitor is a typical color computer monitor connected to a keyboard and pointing devices, all of which are used to manage the images displayed on the two high-resolution monitors, including the application of various special postprocessing features, and manage the patient's file.

The image screening area for radiographers (Figure 12-2) might be considered as a *workstation,* but not a *diagnostic* workstation. Display monitors here are much lower resolution (1000 pixels) to save expense, but are capable of making permanent changes to images and saving them into the PAC system.

A *display station* is a computer terminal that is restricted from saving any permanent changes to an image. Since they are not used by a radi-

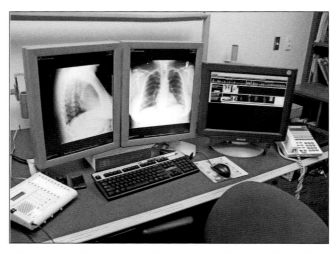

Figure 12-1. A radiologist's *diagnostic workstation* typically consists of two high-resolution (2000-pixel) monitors for image display, and an additional computer terminal with monitor for file management. (From Q. B. Carroll, *Radiography in the Digital Age,* 3rd ed. Springfield, IL: Charles C Thomas, Publisher, Ltd., 2018. Reprinted by permission.)

ologist for diagnosis, display stations typically consist of low-resolution monitors, making them cheap enough to be widely distributed as needed in the emergency department and strategic clinical locations from intensive care units to referring doctors' offices. Although a displayed image can be "windowed" or otherwise adjusted *on the monitor,* changes made cannot be saved into the PAC system.

In this chapter, we will first examine the basic technology behind the *LCD* monitor and how it works. Radiographers should have enough understanding of the LCDs to appreciate their proper *use* and *care.*

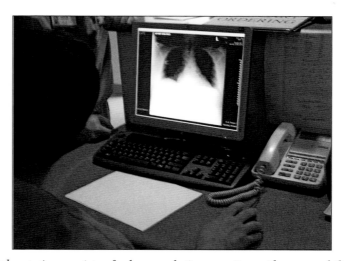

Figure 12-2. A typical *display station* consists of a low-resolution monitor with no capability for making permanent changes to images. (From Q. B. Carroll, *Radiography in the Digital Age,* 3rd ed. Springfield, IL: Charles C Thomas, Publisher, Ltd., 2018. Reprinted by permission.)

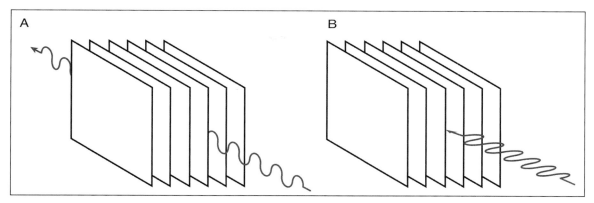

Figure 12-3. In a polarizing lens, only light waves whose electrical component vibrates *parallel* to its long molecular chains can pass through, **A**. Light waves whose electrical component vibrates perpendicular to the iodine chain molecules are blocked, **B**. (From Q. B. Carroll, *Radiography in the Digital Age,* 3rd ed. Springfield, IL: Charles C Thomas, Publisher, Ltd., 2018. Reprinted by permission.)

The Liquid Crystal Display (LCD) Monitor

To understand the LCD, it will be first necessary to overview the concept of *light polarization.* The lenses of polarizing sunglasses contain long chains of iodine molecules, all aligned parallel to each other much like the grid strips in a radiographic grid. Like an x-ray, a *ray of light* is electromagnetic radiation that consists of a double-wave with its electrical component perpendicular to its magnetic component. Shown in Figure 12-3, if the *electrical component* of a light wave is oscillating up and down, it can pass through a polarizing lens with *vertical* string molecules, but *not* through one with *horizontal* string molecules. A light wave will be blocked by string molecules that are *perpendicular* to the electrical wave.

Now, imagine layering two polarizing lenses together such that their string molecules are *perpendicular to each other,* as in Figure 12-4. By doing this, *all* light will be blocked by the compound lens. Any light waves that make it through the first lens will be perpendicular to, and blocked by, the second lens. Shown in Figure 12-5, this is just the arrangement used to construct an LCD monitor screen. It consists of two thin sheets of polarizing glass placed with their iodine chain molecules perpendicular to each other, such that in their default condition, all light is blocked, Figure 12-5**B**.

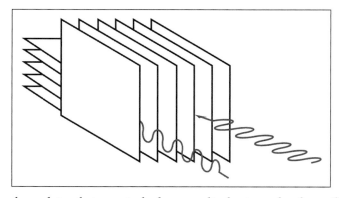

Figure 12-4. Two polarizing lens plates that are stacked perpendicular to each other will block all light from passing through. (From Q. B. Carroll, *Radiography in the Digital Age,* 3rd ed. Springfield, IL: Charles C Thomas, Publisher, Ltd., 2018. Reprinted by permission.)

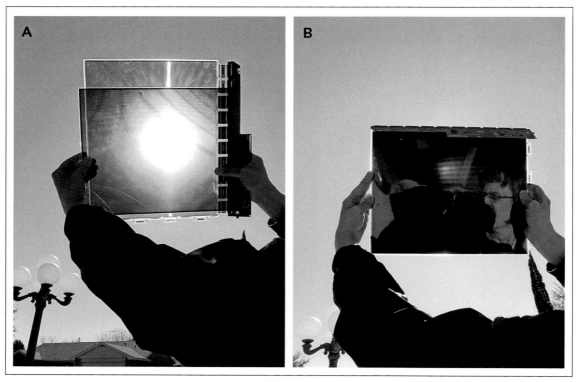

Figure 12-5. The two glass plates form an LCD monitor screen contain light polarizing layers that are perpendicular to each other. Here, when held crosswise to each other, *A*, light from the sun is able to pass through. When held in their normal position, *B*, all light is completely blocked by the perpendicular polarizing lenses. (From Q. B. Carroll, *The Electronic Image: Display and Quality* (video), Denton, TX: Digital Imaging Consultants, 2017. Reprinted by permission.)

In between the two polarizing lenses there is a thin layer of nematic liquid crystal material, Figure 12-6, usually hydrogenated amorphous silicon. The term nematic refers to crystals that have a long linear shape and tend to align parallel to each other (Figure 12-7). Even though hydrogenated amorphous silicon is a crystalline substance, the crystals can slide around each other and flow like a liquid.

Each of the polarizing glass plates that make up the LCD screen includes a layer of thin, flat wires to conduct electricity. Usually made of indium tin oxide, these flat wires are also *transparent* so that light can pass through them. When you touch the surface of an LCD screen, you are directly touching the top layer of these wires, and with excessive pressure they can be damaged. (Most "dead" pixels occur from this type of abuse.) Just as with the polarizing iodine mol-

ecules, the transparent conductors are also aligned *perpendicular* to each other when the two plates are in their normal juxtaposition, effectively forming horizontal *rows* of wires in one sheet and vertical *columns* of wires in the other, such that the conductors crisscross each other across the screen.

Shown in Figure 12-8, each *junction* of these rows and columns of transparent wires forms a single *hardware pixel* in the monitor screen. At each junction, the surfaces of the two wires facing each other constitute two electrodes that can vary the electrical charge between them passing through the nematic liquid crystal layer. The surfaces of the electrodes that are in contact with the nematic liquid crystal are coated with a thin polymer layer that has been rubbed in a single direction with an abrasive cloth material to create a finely *scratched* surface (Figure 12-9).

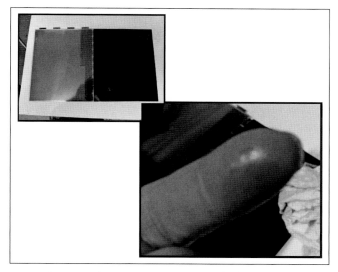

Figure 12-6. Unfolding the two polarizing glass plates of an LCD (*top*), one can rub the nematic liquid crystal off of the plates with a finger. (From Q. B. Carroll, *The Electronic Image: Display and Quality* (video), Denton, TX: Digital Imaging Consultants, 2017. Reprinted by permission.)

Figure 12-7. Nematic crystals are long, linear molecules that tend to align parallel to each other.

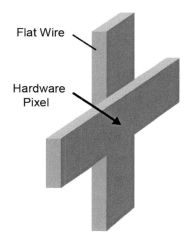

Figure 12-8. On an LCD monitor screen, each *hardware pixel* consists of the junction between two thin, flat wires that act as electrical conductors, but are also transparent to allow light to pass through. (From Q. B. Carroll, *The Electronic Image: Display and Quality* (video), Denton, TX: Digital Imaging Consultants, 2017. Reprinted by permission.)

The scratches on the two opposing surfaces are aligned perpendicular to each other with the nematic liquid crystal layer between them.

The nematic crystals of the hydrogenated amorphous silicon tend to line up with the scratches in the electrode surfaces when *no electric charge is present* and the pixel is in its default "on" state. Since the scratches are perpendicular to each other, the liquid crystals line up in a *spiral* pattern that twists 90 degrees between the two glass plates as shown in Figure 12-9. Light waves passing through the nematic liquid crystal tend to follow the orientation of the crystals. Therefore, the light itself twists 90 degrees between the two plates such that it can pass through the second plate to the observer (Figure 12-10).

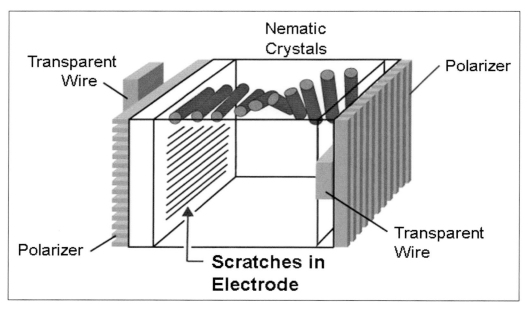

Figure 12-9. In an LCD hardware pixel, when no electric charge is present, the nematic crystals tend to line up with fine scratches etched into the electrode plates. However, scratches in the back plate are perpendicular to those in the front plate. This results in the crystals twisting 90 degrees between the two plates. (From Q. B. Carroll, *The Electronic Image: Display and Quality* (video), Denton, TX: Digital Imaging Consultants, 2017. Reprinted by permission.)

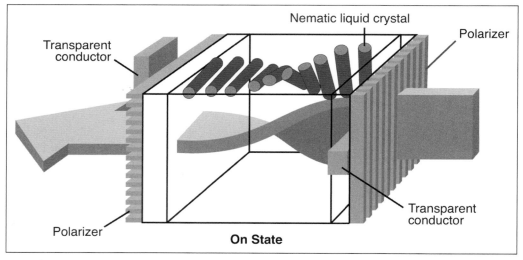

Figure 12-10. When an LCD pixel is in the "on" state, with *no electric charge* applied, light will be twisted along with the orientation of the nematic crystals, allowing the light to pass through both polarizing plates and on to the viewer. (From Q. B. Carroll, *Radiography in the Digital Age,* 3rd ed. Springfield, IL: Charles C Thomas, Publisher, Ltd., 2018. Reprinted by permission.)

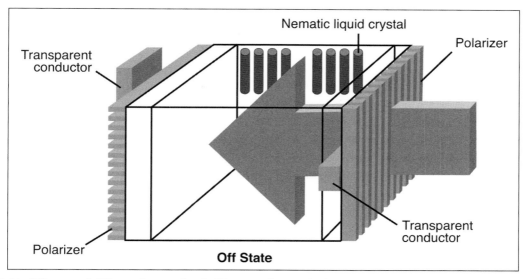

Figure 12-11. When an LCD pixel is in the "off" state, electrical charge is being applied to the electrodes. The nematic crystals align to the electric charge, all pointing the same direction. Light passing through remains "straight" or untwisted and is blocked by the second (front) polarizing plate. (From Q. B. Carroll, *Radiography in the Digital Age,* 3rd ed. Springfield, IL: Charles C Thomas, Publisher, Ltd., 2018. Reprinted by permission.)

Ironically, this "on" state for a pixel is its default state when *no electrical current is applied.* Electrical current is used to turn the pixel *off* as shown in Figure 12-11. When a full electrical charge is applied to the pixel electrodes, the nematic crystals all align themselves in the same direction, parallel to each other, according to the electrical charge present. This places them perpendicular to the second polarizing lens. Light waves following these crystals will be perpendicular to the second lens and blocked by it, leaving a dark spot on the visible screen.

By varying the amount of electrical charge applied to the pixel, we can change the un-twisting effect on the nematic crystals, and the un-twisting of the light, to varying degrees. Thus, a very slight charge will only untwist the crystals slightly, allowing most light to pass through and resulting in a light gray spot. More charge causes a more severe untwisting effect, blocking more light to result in a darker gray spot. A full amount of charge results in a "black" spot where all light has been blocked. The voltage (or amperage) applied to each pixel in each row is controlled by electronic circuits alongside the glass plates (Figure 12-12). On very large monitor screens,

the display must be *multiplexed,* grouping the electrodes in different parts of the screen with a separate voltage source for each group, in order to ensure sufficient voltage to all pixels.

This is the basic process for building up an image on any LCD screen. LCD wristwatches, calculators and other instruments have a familiar gray-silver appearance to them because they use a simple reflective surface behind the polarizing lenses to reflect incoming light. For computer and imaging monitors, a much brighter image is needed, so a source of active *backlighting* is used. This lighting is usually provided by light-emitting diodes (LEDs) or by fluorescent light bulbs. Figure 12-13 demonstrates the most common arrangement: a pair of very thin fluorescent bulbs placed along one side of the LCD screen. By using a whole series of plastic light-diffusing filters shown in Figure 12-14, the light from these two long bulbs can be scattered so evenly across the back of the screen that the brightness of the opposite side of the monitor is only a few percent less than the brightness near the bulbs.

Compared to the slow-responding *passive-matrix LCDs* used for wristwatches and calcula-

Digital Radiography in Practice

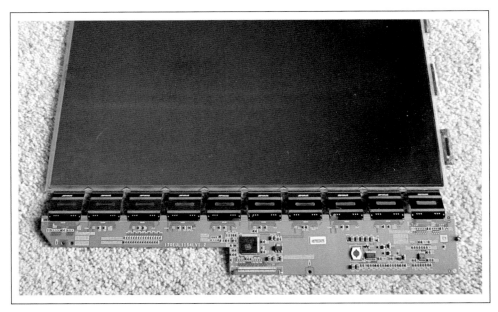

Figure 12-12. Electronic circuits alongside the glass plates of the LCD monitor control the voltage supplied to each pixel in each row. The greater the voltage, the more the nematic crystals are *untwisted,* and the darker the pixel. (From Q. B. Carroll, *The Electronic Image: Display and Quality* (video), Denton, TX: Digital Imaging Consultants, 2017. Reprinted by permission.)

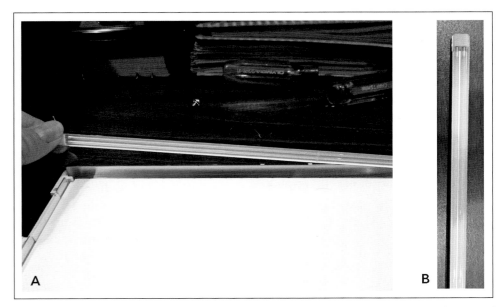

Figure 12-13. A pair of thin fluorescent light bulbs supply this LCD monitor with all the light it needs from a single side of the screen, *A*. Both bulbs can be better seen in a close-up photo, *B*. (From Q. B. Carroll, *The Electronic Image: Display and Quality* (video). Denton, TX: Digital Imaging Consultants, 2017. Reprinted by permission.)

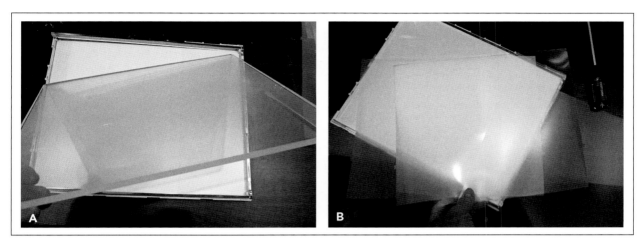

Figure 12-14. Several plastic light-diffusing filters scatter the light from the fluorescent bulbs evenly across the entire surface of the LCD monitor screen. The thickest filter, *A*, accounts for most of the thickness of the LCD. (From Q. B. Carroll, *The Electronic Image: Display and Quality* (video), Denton, TX: Digital Imaging Consultants, 2017. Reprinted by permission.)

tors, much faster response and refresh times are necessary for computer and imaging monitors. The *response time* is the time necessary for the monitor to change its brightness. The *refresh time* is the time required by the entire monitor screen to reconstruct the next frame of a dynamic (moving) image or the next "slide" in a series. To achieve these speeds, an *active-matrix LCD (AMLCD)* is necessary, in which each hardware pixel possesses its own thin-film transistor (TFT) (see Chapter 10). This electronic arrangement allows each column line to access a pixel simultaneously so that entire *rows* of pixels can be read out one at a time, rather than single pixels being read out one at a time. AMLCD monitors can have refresh rates as high as 240 hertz (cycles per second).

Various digital processing operations (described in Chapters 6 and 7) can be applied *by the monitor,* including rescaling, noise reduction, edge enhancement, and the insertion of interpolated frames to smooth out motion, but too much processing can cause unacceptable *input lag* between the time the monitor receives the image and actual display.

Truly "dead" pixels appear as permanent *white* spots, damaged in such a way that electrical charge cannot be applied across the pixel. Permanent *dark* spots on the monitor screen are stuck pixels that are constantly receiving electrical charge. Pixels that are dead or stuck can sometimes be fixed by very gently massaging them with the fingertip. However, remember that hardware pixels can also be damaged by sharp pressure such as from fingernails, so this procedure should not be attempted with very expensive radiologist workstation monitors.

Manufacturers recommend that class 1 LCDs used by radiologists for diagnosis be "nearly perfect." A display monitor should be replaced if there are more than 15 defective pixels across the entire screen, more than 3 defective pixels in any one-centimeter circumference, or more than 3 defective pixels adjacent to each other anywhere on the screen.

LCD Image Quality

The LCD monitor has nearly universally replaced both the older cathode-ray tube (CRT) which was 35-40 cm (14-16 inches) deep from front to back, and the "viewbox" illuminator used to hang hard-copy x-ray film on to view radiographs. The much thinner and lighter LCD has many advantages over these older technologies but also some notable disadvantages.

In the "plus" column, the LCD has perfect geometry so there is no distortion of the image or change in sharpness across different areas of the screen. There is no glare and reflection of ambient light off the screen surface is minimal. There is no flicker and the brightness is almost perfectly uniform and consistent over time.

On the other hand, close inspection reveals *pixelation* of the image in which the dark lines between individual pixels are visible. (Conventional film radiographs had no pixelation.) Contrast is substantially limited to a ratio of about 600:1, whereas the contrast ratio of a viewbox or CRT could be as high as 3000:1. This low contrast is partly caused by the LCD's inability to produce a "true black," as can be observed in a completely darkened room—there is always some light leakage from the backlighting elements in the LCD. As a result, the radiologist must employ frequent "windowing" to optimize each image displayed.

The LCD is sensitive to significant changes in temperature, and requires at least 15 minutes of warm-up time to reach its full luminance after being turned on. Perhaps the most bothersome disadvantage of the LCD is its *viewing angle dependence*—a rapid drop-off of brightness as the viewing angle of the observer is increased. Because of this, the LCD monitor must be viewed almost directly "head-on," that is, perpendicular to the screen, for full visibility. This makes it difficult for secondary observers to get a good view of the monitor image.

Spatial Resolution (Sharpness)

As described in Chapter 8, the sharpness of any radiographic image is ultimately determined by the size of the dexels (dels) used to record the latent image or to the size of the pixels used to process the image. Any effect that

field of view (FOV) or matrix size have on sharpness must be due to their effect on pixel or dexel size. There are different kinds of matrices and fields of view. Depending on how we define the matrix or the FOV, they may or may not affect sharpness. If pixel or dexel size is unchanged, then spatial resolution is unchanged.

For the display monitor, we are concerned with *two* types of pixel: the *hardware pixels* of the display monitor itself (which determine the monitor's *inherent resolution* capability), and the "soft" pixels of the actual displayed light image, which are relative and can be changed by zooming the field of view in or out. In this chapter, we are primarily concerned with the inherent resolution of the display monitor. This is a fixed value, dependent upon the size of the *hardware pixels*, the intersections between the flat conducting wires in the monitor screen. The smaller these pixels are, the better the sharpness of the monitor and the higher its inherent spatial resolution.

Dot pitch or *pixel pitch* is the distance between the centers of any two adjacent hardware pixels, Figure 12-15. The smaller the pixel pitch, the smaller the pixels themselves must be, and the sharper the spatial resolution. We often describe the resolution of a camera or a display monitor by referring to its total number of pixels, a "3-megapixel camera" or a "5-megapixel monitor." In popular use, the higher this total number of pixels, the higher the resolution.

But, keep in mind that all this *assumes a fixed physical detection or display area*. In other words, if we are comparing two display monitors that both have 35 cm (14-inch) screens, then it would be true that a 5-megapixel monitor would have better resolution than a 4-megapixel monitor because the hardware pixels would have to be smaller to fit into the physical size of the screen. But, if one monitor is 35 cm (14 inches) and another is 45 cm (18 inches), the larger monitor could have hundreds more pixels *of the same size*. In this case, their resolution would be equal.

Generally, the pixel size on a computer monitor is about 0.2 mm and for a high-resolution Class 1 diagnostic monitor it can be as small as

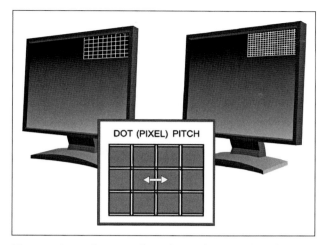

Figure 12-15. On an LCD, *dot pitch* or *pixel pitch* is the distance between the centers of two adjacent hardware pixels. For a fixed screen size, the smaller the pixel pitch (*right*), the larger the matrix and the higher the spatial resolution. A 5-megapixel (matrix size) monitor is sharper than a 3-megapixel monitor. (*Courtesy,* Digital Imaging Consultants, Denton, TX. Reprinted by permission.)

0.1 mm. Remember that typical focal spot sizes are 0.5–0.6 mm for the small *focal spot* and 1.0–1.2 mm for the large focal spot. Because of the large ratio between the source-to-object distance (SOD) and the object-to-image receptor distance (OID) used in standard radiography, the width of the penumbra is reduced to a fraction that can be roughly one-tenth of the focal spot size. Note that for a large focal spot size of 1.2 mm, this would come to 0.12 mm of penumbral blur at the edge of a detail. This rivals the 0.1 mm pixel size for a Class 1 diagnostic monitor. What this means is that although the display monitor is generally the *weakest link* in the imaging chain for image sharpness, *it is possible, especially when using the large focal spot, for the focal spot to be causing more blur than the display monitor* at least in some cases. Many radiographers have developed the very poor habit of using the large focal spot on procedures, such as the distal extremities, that should be done with the *small* focal spot. In the digital age, this poor practice could still result in images that are more blurry than necessary.

Luminance and Contrast

Because of the inherently limited contrast capability of the LCD already noted, it is essential that viewing conditions be optimized and that the brightness of the LCD monitor itself be carefully monitored. In this relation, we introduce three units of brightness with which the radiographer should be familiar: the *candela,* the *lumen,* and the *lux.* Several other units are also used, but these three are the most common and will suffice for our purposes.

Luminance refers to the rate of light (brightness) emitted from a source such as an LCD monitor. Light emitted from the display monitor spreads out in all directions within a dome-shaped hemisphere, as shown in Figure 12-16. To obtain a measure of brightness, we must define a specific area within this "dome" in which to take our measurement. We divide the volume of space within this hemisphere into units called *steradians.*

Shown in Figure 12-16, one steradian is a *cone* with a radius r from the source of light, whose base has an area equal to *r-squared (r^2).* This formula results in a specific angle that sweeps out the cone in space, such that approximately 12.5 steradians always fit within any sphere. Thus, for the *hemisphere* of light projected in front of an LCD monitor, there are just over 6 steradians formed. This gives us a fixed, standardized area, r^2, the base of one of these cones, in which to measure and compare brightness.

Now, in terms of total energy, the amount of power emitted from a typical candle is 0.0015 watts per steradian. (Remember that the term *watt* represents the Joules of energy spent *per second,* so that time is already taken into account in this formula.) We define this amount of brightness as *one candela,* abbreviated *Cd.* Strictly speaking, the *candela* unit describes the rate of light emitted *in all directions* from a candle, or the power of the candle itself, whereas the unit *lumen* refers to the brightness of that light within one single steradian. In other words, a light *source* with one candela of brightness generates one lumen of light *flow* through each steradian of

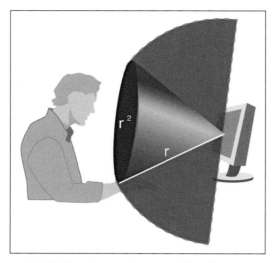

Figure 12-16. *Luminance* is the rate of brightness emitted from the display monitor. One candela of overall light emission is defined as one *lumen* per steradian. A *steradian* is defined as a cone whose base has an area equal to the square of the radius (distance) from the light source. The steradian is about one-sixth of a hemisphere. (From Q. B. Carroll, *Radiography in the Digital Age,* 3rd ed. Springfield, IL: Charles C Thomas, Publisher, Ltd., 2018. Reprinted by permission.)

space around the source. Expressed as a formula:

$$1 \text{ Cd} \rightarrow 1 \text{ Lm/sr}$$

where *sr* refers to the steradian.

The *photometer,* Figure 12-17, is the common device used to measure the brightness output of an LCD or other display monitor. Using the photoelectric effect, this device converts light photon energy into electrical charge or current to generate a display readout. The readout can be in *lumens* or in *candela per square meter (Cd/m²)*. (A similar device called a *densitometer* is used to measure the degree of "blackness" on a hard-copy printout of a radiograph. It mathematically inverts the brightness measurement to indicate how much light is being *blocked* by a particular area on the hard-copy image.)

The American College of Radiology (ACR) requires a minimum brightness capability for radiologic display systems of 250 lumens. Typical brightness settings preferred by radiologists

fall in the range of 500–600 Lm. The maximum luminance capability of most LCDs is about 800 Lm or 800 Cd/m².

Optimum viewing of diagnostic LCD images requires that the ambient reading room light be dimmed. This avoids reflectance of ambient light off the surface of the display monitor screen which is destructive to visual image contrast. *Specular reflectance* refers to the reflection of actual light sources such as a light bulb or window; *diffuse reflectance* is the cumulative effect of room lighting across the area of the monitor screen.

The term *illuminance* refers to the rate of light striking a surface. (When reading a book, *luminance* would refer to the brightness of a lamp behind you, whereas *illuminance* would describe how bright the pages of the actual book appear to you.) The effect of ambient room lighting on the surface of an LCD monitor screen, including reflectance off the screen, and the resulting reduction in visible contrast of the image, is a result of *illuminance.* The most common unit for illuminance is the lux, defined as one lumen per square meter of surface area.

On a sunny day outdoors, about 105 lux illuminates the sidewalk in front of you. At night, a full moon provides about 10 lux of visibility.

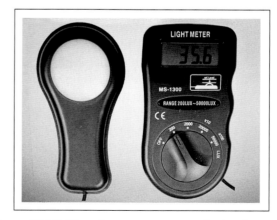

Figure 12-17. The *photometer* measures the brightness of light output from a display monitor. Readout units can be in *lux, lumens* or *candela per square meter.* (From Q. B. Carroll, *Radiography · in the Digital Age,* 3rd ed. Springfield, IL: Charles C Thomas, Publisher, Ltd., 2018. Reprinted by permission.)

Typical indoor office lighting ranges from 75 to 100 lux. For accurate medical diagnosis, *the ambient lighting for a radiologists' reading room should never exceed one-quarter of normal office lighting, or 25 lux.*

We've listed *viewing angle dependence* as a major disadvantage of the LCD. Viewing angle dependence can be directly measured and used to compare different types of LCD monitors. To do this, a photometer can be placed at a precise distance perpendicular to the center of the LCD monitor for a base measurement, then measurements can be repeated at increasing angles from the perpendicular while carefully maintaining the distance. Each of these measurements is divided by the original perpendicular measurement to obtain a percentage. More expensive LCDs may have every other column of pixels "canted" or slightly tilted to reduce viewing angle effects.

Nature of Display Pixels

When it comes to the display monitor, each *hardware* pixel possesses a definite shape (roughly square) and a definite area, about the size of a 10-point font period, just a bit smaller than the period at the end of this sentence. In a *color* monitor, each of these pixels must be capable of displaying the entire range of colors in the spectrum. Shown in Figure 12-18, this is achieved by using three rectangular *subpixels*—one red, one blue, and one yellow-green—energized at different combinations of brightness. When all three subpixels are lit together, the result of this mixing is white light. It is necessary, then, that we distinguish between what constitutes a *pixel* and what constitutes a *subpixel.*

In constructing a digital image for display, we define a *pixel* as the *smallest screen element that can display all gray levels or colors within the dynamic range* of the system. Thus, for a color monitor, each subpixel can only display red, blue, or yellow-green, but only the entire pixel, consisting of all three subpixels working in tandem, can display the full range of colors including white.

Monochrome or "black-and-white" display monitors are sufficient for most medical imag-

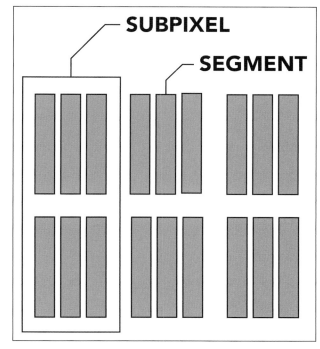

Figure 12-18. A typical hardware pixel for an LCD is composed of 18 phosphor segments arranged into *domains* of 3 each and *subpixels* of 6 each. In a monochrome (black and white) monitor, each subpixel can be addressed as a separate pixel, tripling resolution. (From Q. B. Carroll, *Radiography in the Digital Age,* 3rd ed. Springfield, IL: Charles C Thomas, Publisher, Ltd., 2018. Reprinted by permission.)

ing purposes. Figure 12-18 illustrates how each hardware pixel actually consists of 18 individual segments that can be illuminated. These segments are arranged in groups of three called *domains,* and a pair of domains (or 6 segments) constitutes one *subpixel.* For a monochrome display monitor, each subpixel is individually addressed by the computer, and so each subpixel has the capability of displaying the full range of gray shades from black to white. This meets our definition of a pixel. In effect, for a monochrome monitor, each subpixel acts as a full pixel at one-third the size of a color monitor pixel, and three times the sharpness can be achieved. This enhanced sharpness is a great advantage of monochrome display monitors in medical imaging.

Chapter Review Questions

1. Two *Class 1* high-resolution monochrome display monitors are required for a radiologist's diagnostic _____.

2. A computer terminal that is restricted from saving any permanent changes to images into the PAC system is classified as a _____ station.

3. A light wave will be blocked by string iodine molecules that are oriented _____ to the electrical component of the light wave.

4. In its normal configuration, a computer monitor screen consists of two thin sheets of polarizing glass with their iodine molecules oriented _____ to each other.

5. Crystalline molecules that have a long, linear shape and tend to align parallel to each other are called _____ crystals.

6. Each hardware pixel of an LCD monitor screen consists of two flat, transparent _____ crossing over each other.

7. Because the touchable surface of the LCD screen is made of these pixels, applying excessive pressure with the fingers can _____ them.

8. When no electric charge is present, the crystals between the two polarizing sheets line up in a *spiral* pattern because fine _____ in the electrodes on either side are perpendicular to each other.

9. Light passing through the LCD screen tends to follow the orientation of the liquid _____.

10. When an electric charge is applied to the pixel, all of the crystals line up _____ to each other, and the *second* polarizing sheet blocks them.

11. The degree of molecular *twisting,* and therefore the amount of light passing through the LCD screen, is controlled by the amount of _____ applied to the pixel.

12. Necessary backlighting for the AMLCD is provided by _____ or by fluorescent _____.

13. To evenly spread the light across the screen area, a series of light-diffusing _____ are used.

14. For an LCD, define *response time:*

15. For an LCD, define *refresh time:*

16. To allow whole rows of pixels to be read out at one time for higher speed, in an _____ _____ LCD each column line is able to access a pixel simultaneously from all of the rows.

17. A truly "dead" pixel appears as a permanent _____ spot.

18. Due to their backlighting, LCD monitors have limited contrast because of their inability to produce a "true-_____."

19. When multiple observers want to see an image, the greatest disadvantage for LCD monitors is their *viewing _____ dependence.*

20. *Assuming they have the same physical display area,* a 5-megapixel monitor will have _____ sharpness than a 3-megapixel monitor.

21. The hardware pixel size for display monitors ranges from ____ to ____ millimeters.

22. Although the display monitor is usually the limiting factor for image sharpness, it is *possible* in some positioning circumstances for the *large* focal spot to cause more _____ than the monitor.

23. The rate of light emitted from a display monitor defines its _____.

24. The photometer measures the rate of emitted light usually in units of _____, _____, or _____ ____ ____ _____.

25. Reflectance of ambient room light off the display screen surface has the effect of reducing the apparent _____ of the image.

26. The illuminance in a radiologic reading room should never exceed _____ of normal office lighting.

27. For a display monitor, we define a hardware *pixel* as the *smallest screen element capable of displaying all of the _____ levels in the dynamic range of the system.*

28. For an LCD, a single hardware pixel is actually made up of three rectangular _____, each of which can be individually addressed by the computer.

Chapter 13

ARCHIVING PATIENT IMAGES AND INFORMATION

Objectives

Upon completion of this chapter, you should be able to:

1. Define the main components and functions of a picture archiving and communication system (PACS).
2. Define *metadata* and *DICOM header*.
3. State the DICOM and HL-7 standards and their purpose
4. Describe the major concerns for image compression, storage, security, and transmission and how they are addressed.
5. Define *medical imaging informatics* and distinguish between *EMRs* and *EHRs*.
6. List the three major components of security for an information system
7. Describe the physical connections and bandwidth between nodes of an information system.

Picture Archiving and Communication Systems (PACS)

Through digitization, images from all the different modalities within a medical imaging department (DR, CR, CT, MRI, DF, NM, sonography, and so on) can be stored on optical discs, magnetic discs or magnetic tapes for universal access through a *picture archiving and communications system* or *PACS*. The PAC system provides long-term archival storage, allows retrieval of the images for viewing on television-type monitors, and allows transmission of the images to remote clinical sites and hospitals.

Image acquisition stations in clinics and hospital departments that use the PAC system are considered as *service class users (SCUs)*, while the centralized devices that manage storage and distribution of the images, along with PACS workstations for accessing and manipulating the images, are considered as *service class providers (SCPs)*.

Figure 13-1 diagrams the components of a typical PAC system. We see that a PACS is actually a type of *local area network* or *LAN*. This LAN is tasked with managing the flow of digital images, related metadata and patient information from the various imaging modalities through a central *control computer* to the radiology information system (RIS), the hospital information system (HIS), and workstations throughout the network. These workstations can be in hospital departments such as the ER or CCU, in physician clinics, or in the homes of radiologists.

Within the central control computer, or *PACS server* (Figure 13-2), images are stored en masse using magnetic or optical jukeboxes, stacks of over 100 magnetic or optical discs. A modern PACS can store over 1 million medical images. Newer systems use *fiber optic* technology that can transmit data at much higher speeds than electronic transmission lines. Some systems employ a *wide area network (WAN)* to allow access to images and data at remote locations.

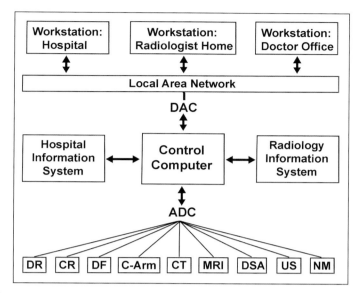

Figure 13-1. Diagram of a PACS. The control computer receives acquired images and patient information from all digitized imaging modalities. Through a local area network (LAN), these images and information are made available to workstations and display stations throughout the hospital, affiliated clinics and radiologists' homes, along with the hospital and radiology information systems (HIS and RIS).

In the early days of PAC systems different manufacturers used different computer languages and protocols that were not compatible with each other and impeded communication between clinics and hospitals at the expense of patient care. It was quickly recognized that industry-wide standardization was needed. Therefore, in the 1980s, the American College of Radiology (ACR) and the National Electronic Manufacturers Association (NEMA) joined to formulate the *Digital Imaging and Communication in Medicine (DICOM)* standard. While still allowing manufacturers to use different architecture and even terminology between their systems, DICOM standardizes the *transmission behavior* of all the devices used in various PAC systems. Patient care has greatly benefitted from the improved efficiency of communication throughout the healthcare system.

The essential purpose of the PAC system is to act as a *database.* The status of any patient or radiologic study can be displayed as pending, in progress, or completed. The user can initiate

Figure 13-2. A PACS Server. (From Q. B. Carroll, *Radiography in the Digital Age,* 3rd ed. Springfield, IL: Charles C Thomas, Publisher, Ltd., 2018. Reprinted by permission.)

	Tag ID	Description	Value
A	(0008,0021)	Series Date	20130709
	(0008,0022)	Acquisition Date	20130709
	(0008,0030)	Study Time	171352.000000
	(0008,0031)	Series Time	171411.000000
	(0008,0032)	Acquisition Time	171519.000000
	(0008,0050)	**Accession Number**	**884279001**
	(0008,0060)	Modality	DX
	(0008,0068)	Presentation Intent Type	FOR PRESENTATION
	(0008,0070)	Manufacturer	Canon, Inc.
	(0008,0080)	Institution Name	
	(0008,0090)	Referring Physician's Name	
	(0008,1010)	Station Name	CXDI-5821597MA
	(0008,1030)	Study Description	
	(0008,103E)	Series Description	CXR PA

	Tag ID	Description	Value
B	**(0010,0010)**	**Patient's Name**	**John Doe**
	(0010,0020)	**Patient ID**	**139967010**
	(0010,0030)	Patient's Birth Date	19580105
	(0010,0040)	Patient's Sex	M
	(0010,0010)	Patient's Age	57
	(0018,0000)	Unknown	338
	(0018,0015)	**Body Part Examined**	**CHEST**
	(0018,1508)	Positioner Type	
	(0018,5101)	**View Position**	**PA**
	(0020,0010)	Study ID	51080.1
	(0020,0020)	**Patient Orientation**	**L\F**
	(0020,0062)	Image Laterality	U

	Tag ID	Description	Value
C	(0018,0060)	KVP	+80.0
	(0018,0000)	Device Serial Number	10000932
	(0018,1020)	Software Version(s)	V6.00.15
	(0018,1050)	Spatial Resolution	0.160
	(0018,1052)	Exposure	+32
	(0018,1064)	Imager Pixel Spacing	0.160\0.160
	(0018,1180)	Collimator/grid Name	119cm/8:1/40/Al
	(0018,1200)	Date of Last Calibration	20131107
	(0018,1201)	Time of Last Calibration	110113.000000
	(0018,1401)	**Acquisition Device Processing Code**	**REX454Q2W3LP3GSS17,20Z1M2,1,3P1**
	(0018,1405)	**Relative X-ray Exposure**	**408**

Figure 13-3A. Key types of metadata stored in the DICOM header file are bolded in these excerpts taken from an actual header. Section *A* presents initial identification of the exam (including accession number), equipment, institution, and physicians. Section *B* includes patient ID and demographic information, and details of the exam. Section *C* shows specifics on the radiographic technique used, exposure delivered, and state of equipment. (From Q. B. Carroll, *Radiography in the Digital Age,* 3rd ed. Springfield, IL: Charles C Thomas, Publisher, Ltd., 2018. Reprinted by permission.)

searches for any particular patient, exam, or image. The command *DICOM query/retrieve* searches out a specific image. *DICOM get worklist* imports patient information and study requisitions from the RIS or HIS. *DICOM send* conveys data to the general network.

Attached to every image is a *DICOM header,* a summary of critical identification and study information such as the date and time of the procedure and the number of images taken, essential information on the image itself such as the body part, position, technique used, image format and receptor size, and the parameters used to digitally process the image, along with where additional information associated with the image can be accessed. The DICOM header, which should appear in a bar at the top of the screen at the push of a button for each image, is

	Tag ID	Description	Value
D	(0018,7004)	Detector Type	SCINTILLATOR
	(0018,7005)	Detector Configuration	AREA
	(0018,700A)	Detector ID	7e00182
	(0018,7030)	Field of View Origin	352\0
	(0018,7032)	Field of View Rotation	0
	(0018,7034)	Field of View Horizontal Flip	YES
	(0019,0016)	Unknown	Canon, Inc
	(0020,0011)	Series Number	1
	(0020,0013)	Instance Number	1

	Tag ID	Description	Value
E	(0028,0002)	Samples Per Pixel	1
	(0028,0004)	Photometric Interpretation	MONOCHROME2
	(0028,0010)	**Rows**	**2688**
	(0028,0011)	**Columns**	**2128**
	(0028,0100)	**Bits Allocated**	**16**
	(0028,0101)	Bits Stored	12
	(0028,0102)	High Bit	11
	(0028,0103)	Pixel Representation	0
	(0028,0301)	Burned on Annotation	NO
	(0028,1040)	Pixel Intensity Relationship	LOG
	(0028,1041)	Pixel Intensity Relationship Sign	1
	(0028,1050)	**Window Center**	**2048**
	(0028,1051)	**Window Width**	**4096**
	(0028,1052)	**Rescale Intercept**	**0**
	(0028,1053)	**Rescale Slope**	**1**
	(0028,1054)	**Rescale Type**	**US**
	(0028,2110)	Lossy Image Compression	00
	(0040,0555)	Acquisition Context Sequence	
	(2050,0020)	**Presentation LUT Shape**	**IDENTITY**
	(7ZFE,0010)	Pixel Data	00 00 00 00 00 ...

Figure 13-3B. Metadata stored in the DICOM header, continued: Section **D** provides detailed information on the detector used and field orientation. Section **E** lists the parameters used for digital processing, including matrix size, window level and window width, and the type of rescaling curve and look-up tables (LUTs) applied. (From Q. B. Carroll, *Radiography in the Digital Age,* 3rd ed. Springfield, IL: Charles C Thomas, Publisher, Ltd., 2018. Reprinted by permission.)

a *summary of essential metadata* for the image. The term *metadata* refers to all of the associated database information "behind" each image, and includes 80 to 100 lines of details as shown in Figure 13-3.

At a PAC workstation any image can be modified by windowing its brightness and contrast, by magnification and cropping, by adding various types of annotation, or by applying a number of special features that effectively re-process the image (such as edge-enhancement or smoothing). For radiographers, it must be emphasized that some of the original image data can be lost if image quality is altered and then the changed image is *saved into the PACS* without first saving a copy of the original. This loss of data can limit the radiologist's ability to window the image as needed, so it is best to always make a copy of the original, then modify and save the copy as an additional image.

The digital storage capacity required of PAC systems can become so extreme that some degree of *image compression* is necessitated. A single DR or CR image requires 4–5 megabytes of storage space. In a typical imaging department taking 200 such images per day, this comes to 30 *gigabytes (GB)* of storage needed per month just for DR and CR. MRI scans require *3 times* more storage per image, and CT scans require up to *5 times* more per image, such that these four diagnostic modalities together can easily exceed 100 gigabytes per month of needed storage. Cost-effectiveness demands that image file sizes be compressed as much as feasible.

To compress an image file, several adjacent microscopic pixels must be combined to effectively form a larger pixel with an averaged value. When this operation is carried out globally across the entire image, a smaller matrix with fewer pixels results. Fewer pixel values need be stored by the computer, saving storage space. But, there are two negative side-effects: First, since the image pixels are now larger in size, small image details may be lost resulting in less sharpness in the image. Second, image noise will also become more apparent because small defects will be magnified in size. So, there is a trade-off between saved storage space and image quality. A balance must be struck where these negative effects are not severe enough to compromise diagnosis.

The American College of Radiology (ACR) has defined *lossless* compression ratios as those less than 8:1, assuming that the original images are standard high-resolution digital radiographs. These have been generally defined as possessing "acceptable levels" of noise and unsharpness that do not hinder proper diagnosis. The ACR defines *lossy* compression ratios as those above 10:1, that result in an "irreversible loss of detail in the image" that is unacceptable for medical applications.

Image storage is distributed among several nodes within the PAC system. Each *acquisition station* at a DR, CR, CT, or MRI unit has a certain limited storage capacity. Assuming that all useful images will be sent into central PACS storage by the radiographer, acquisition workstations are usually programmed to erase images on a regular basis, such as 24 hours after each acquisition, in order to free up storage space for new studies. The *diagnostic (level 1) workstation* used by a radiologist must have much higher storage capacity because recently acquired images are often compared side-by-side with previous studies that must be immediately available. The *quality assurance workstation* acts as a hub for several acquisition units within a department and must also have higher storage capacity. Diagnostic and QC workstations are both

typically programmed to automatically erase images after 5 days of storage.

The PACS server itself (Figure 13-2) defines any image file that has been recently accessed as "active" and temporarily keeps it in *on-line storage,* which uses optical discs or magnetic tape media. When several days have passed without a file being accessed, it is moved from on-line storage to *archive storage* or long-term storage consisting of dozens of optical or magnetic disc *jukeboxes.* These may be in a separate location or even a separate building from the main server. It can take more than five minutes to retrieve a study from long-term jukebox storage.

Because the radiologist often needs to compare previous studies on a particular patient with a recently acquired study, *prefetching programs* have become widely popular in PAC systems. Prefetching programs work during the night, anticipating the need to compare previous studies for patients scheduled the following day, searching the HIS and RIS for studies and records on these patients and bringing them up into *on-line storage* for immediate access. Different protocols are followed according to the scheduled procedure; For a pre-surgery patient, the prefetching program may automatically make the previous two chest exams available, or for a radiographic study of the spine only the most recent spine series. Immediate access to these prior studies can also be useful to the radiographer as he/she prepares for a particular radiographic procedure.

Other advanced features of the PAC system can be included at the radiologist's workstation. These include intelligent image "hanging" protocols that orient displayed images according to the type of exam and can be customized by individual radiologists, decision support tools such as *computer-aided diagnosis (CAD),* and special digital processing algorithms that can be customized by pathology to be ruled out or by radiologist preference.

PACS images can be integrated into the patient's electronic medical record (see the next section) as part of the HIS, in effect becoming part of the patient's *chart.* This makes them accessible in a user-friendly way since no special training for the PAC system is needed, but the displayed image cannot be windowed or otherwise manipulated except by controls on computer display monitor being used. This can be a good option for internal medicine, family medicine, and other general outpatient clinics tied into the hospital system.

Cardiologist and orthopedic surgeon offices will want to be able to fully manipulate the radiographic images. PACS software can be installed on PC's at their clinics to empower most features with only a few limitations. The highest viewing quality and most powerful options require actual PACS workstations at the facility or in the hospital department. Special viewing tools allow interactive consultation between physicians who can be viewing the images simultaneously.

Radiographic images can be burned onto a DVD, along with reports, so a patient can take them to a referring doctor. However, at the receiving end, when these images are displayed on PC monitors that were not specifically designed for radiologic diagnosis, image quality often suffers. A *DICOM viewer* is a software package for displaying the images with the highest resolution and quality possible for a particular display monitor. A DICOM viewer program should always be included on the DVD along with the images themselves. Some basic DICOM viewer programs are available for download on the internet free of charge.

A patient's electronic medical record may also include non-radiographic images such as photographs from a colonoscopy or from surgery. Because of their high resolution, medical images often comprise files too large to send as an attachment to an email, and compression can downgrade image quality enough to compromise diagnosis. Perhaps the best way to send medical images over the internet is to generate a unique *URL (uniform resource locator)* within the PAC system accessible only with a password. An e-mail to a referring physician can include a link to this website address. With the password, the referring physician can then directly access the PACS server holding the files and the DICOM viewer software to properly display them.

Medical images must not be vulnerable to loss through a computer drive "crashing" or to electrical failure in a single system. The *redundant array of independent discs* or *RAID* system was developed to prevent permanent loss of images and records. The RAID system distributes copies of the same data files across several computer hard drives which are independent of each other for power. By spreading these drives across remote geographical locations, even natural disasters such as floods and earthquakes can be prepared for. In a *storage area network (SAN)*, several storage devices are connected in parallel as shown in Figure 13-4. If any one component of this network fails, the redundant connections between the other components are not broken, so the functionality of the overall system is preserved.

Medical Imaging Informatics

Medical informatics is the use of networked computer servers and workstations to improve the efficiency, accuracy, and real-time interactive functionality of medical services. The need for worldwide *standardization* was recognized as early as the 1980s for the way healthcare information is retrieved, integrated and shared. The *Health Level Seven International* committee first met in 1987 for this purpose, and published the *Health Level Seven Standard (HL7)*. HL7 is to medical information and record-keeping what DICOM is to medical images. It allows data storage devices from various different manufacturers to communicate with each other seamlessly.

The heart of medical informatics is the *electronic medical record (EMR)*, a digital version of each patients "chart." The EMR includes physicians' and nurses' notes, laboratory and pathology results, and radiology reports. Many PAC systems now integrate with the hospital information system (HIS) to make all medical images accessible to referring physicians as part of each patient's EMR. An outstanding example of how electronic record-keeping improves the efficiency of patient care is that the system can be programmed to automatically provide timely reminders when screening procedures or preventative check-ups are due.

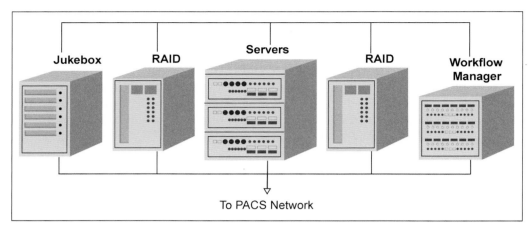

Figure 13-4. A *storage area network (SAN)* is a secure sub-network connecting jukebox storage and a workflow manager to servers and to redundant arrays of independent discs (RAIDs). By connecting all these devices *in parallel,* if any one component fails, the connections between the other components are not broken. (From Q. B. Carroll, *Radiography in the Digital Age,* 3rd ed. Springfield, IL: Charles C Thomas, Publisher, Ltd., 2018. Reprinted by permission.)

Many healthcare providers and insurance companies are now making the entire EMR accessible at any time to the patient via the internet. Healthcare providers are accountable by law to maintain each patient's privacy throughout all their electronic media and record-keeping, including internet access. In the U.S., the *Healthcare Insurance Portability and Accountability Act (HIPAA),* passed in 1996, holds all healthcare providers liable for violations of any patient's right to privacy. This applies not only to the handling and sharing of electronic and printed records, but also to verbal discussions of a patient's condition that might be held within earshot of other people. Radiographers must take special care in these settings.

Authentication programs allow senders and recipients of electronic information to verify each other's identities. Finally, *system integrity* assures that incoming information has not been altered, either accidentally or intentionally. This is the function of a *firewall,* a router, computer, or small network that checks for any history of tampering with incoming files. These, then, are the three basic elements of security for computer communications: *Privacy, authentication,* and *system integrity.*

The *hospital information system (HIS)* integrates the flow of all information within a hospital or clinic. It can be divided into two broad applications: *Clinical information systems (CIS's)* include the radiology information system (RIS), the laboratory information system, and the nursing information system, and others dealing directly with clinical care. *Administrative information systems (AIS's)* manage patient registration, scheduling and billing, and human resources including payroll.

For tracking purposes, the HIS assigns a unique identification number for each patient. For a radiology patient, this number is sent to the image acquisition unit, the PAC system, and the worklist server. When images or records are moved about, the PAC system verifies a match between the ID numbers before storing them. In addition, the RIS assigns a unique number to each particular radiologic procedure or exam, called the *accession number,* which should appear in the DICOM header for every image. The accession number is essential for coordination between radiologists, radiology transcriptionists, radiographers and scheduling and clerical staff. The *RIS/PACS broker* is a device that links textual patient information from the RIS to images from the PACS.

Physical connections between the various nodes of any computer network can be made by optical fiber cables or coaxial cables. Wireless connections can be made by radio waves or by microwaves transmitted between antennas or satellites. The rate at which data can be transferred through any system is referred to as the system's *bandwidth,* and is measured in units of bits per second (bps) or bytes per second (Bps). (Recall that a byte consists of eight bits.)

The exchange of data between computers is governed by transmission control protocols (TCP's) and internet protocols (IP's). A *protocol* is a set of rules that facilitate the exchange of data between the different nodes of a computer network. A *transmission control protocol (TCP)* divides the information to be transferred into "data packets" of a limited size manageable by the transmitting system. *Internet protocol (IP)* formats the actual transmission of the data. *Routers* use these protocols to connect wide-area networks (WANs) to and through the internet. An *intranet* system describes a local area network (LAN) that is used only within a single organization or building. DICOM is the adopted protocol for PAC systems to manage medical images.

Chapter Review Questions

1. What does PACS stand for?
2. Within a PAS, centralized devices that manage storage and distribution of images are considered as *service class* _____.
3. Within a PACS, the central *control computer* can be accessed by workstations in various locations through a _____ _____ *network.*
4. Within the PACS, images are stored en masse using magnetic or optical _____.
5. The *transmission behavior* of all devices used in various PAC systems has been universally standardized by the _____ *standard.*
6. Within the PACS, patient information and study requisitions from the RIS or HIS are imported using the DICOM command *get* _____.
7. At the push of a button, a summary of critical identification and study information for any image will appear in a bar called the *DICOM* _____.
8. All of the database information associated with each image is collectively referred to as _____.
9. When storing thousands of medical images, the sheer volume of information necessitates some degree of _____.
10. A *lossless* image compression ratio has been defined as *less than* _____.
11. Image compression ratios greater than 10:1 are described as _____.
12. Each image acquisition station has limited storage capacity, and so is programmed to _____ images after a designated time period.
13. Recently accessed image files are defined as *active* and are made temporarily available from _____ storage rather than archive storage.
14. A program that automatically accesses older images for comparison in preparation for radiographic studies ordered the next day is called a _____ program.
15. To maintain high image resolution and quality upon display at referring doctors' offices that are not part of the PAC system, *DICOM* _____ software should be included with images burned onto CDs or DVDs.
16. To protect against accidental loss, copies of image files can be distributed across several computer hard drives at different locations. This is called a _____ system.
17. In a *storage area network,* several devices are connected in _____ to protect against natural disasters.
18. Standardization of the exchange of medical information is provided in the publication _____.
19. Within a particular clinic or department, a digital version of a patient's "chart" is referred to as the patient's _____ _____.
20. In the U.S., *HIPAA* standards enforce the privacy of patients' information in all communication.

21. List the three basic elements of security for computer communications:

22. When two users share information through their computers, _____ programs can be used to verify their identities for security.

23. The *radiology information system (RIS)* assigns unique number to each radiologic procedure or exam, called the _____ number.

24. The *rate* at which data can be transferred between devices is the system's _____.

25. Information to be transmitted is divided into manageable "data packets" by a _____ _____ *protocol.*

Chapter 14

DIGITAL FLUOROSCOPY

▬ ▬ ▬ ▬ ▬ ▬ ▬ ▬ ▬ ▬ ▬ ▬ ▬ ▬ ▬ ▬ ▬ ▬ ▬ ▬

Objectives

Upon completion of this chapter, you should be able to:

1. Describe the use and benefits of *dynamic flat panel detectors* for modern fluoroscopy.
2. Give a basic comparison of how the charge-coupled device (CCD) and the complimentary metal-oxide semiconductor (CMOS) work.
3. State the advantages and disadvantages of CCD, CMOS and the conventional electronic image intensifier for fluoroscopic imaging.
4. Explain several ways that digital fluoroscopy equipment can be used to minimize patient dose to x-rays.

Fluoroscopy is the real-time imaging of dynamic events as they occur, in other words, x-ray images in the form of "motion pictures" or "movies." Thomas Edison is credited with inventing the first hand-held fluoroscopic device and coining the phrase "fluoroscope." It consisted of a fluorescent screen mounted at the wide end of a light-tight metal cone and a viewing window at the narrow end. The fluoroscope was held over the patient as the x-ray beam passed through from the opposite side. The remnant x-ray beam exiting the patient struck the fluorescent screen, causing it to glow. For four decades, variations on this device were used that re-

quired a sustained high rate of radiation exposure, often resulting in dangerous levels of radiation dose to patients.

The first electronic *image intensifier* tube was developed by John Coltman in the late 1940s, drastically lowering radiation dose to the patient and making fluoroscopy a mainstream medical practice. By adding a photocathode layer behind the fluorescent screen, light emitted by the screen could be converted into a beam of electrons. Using a series of anodes, these electrons were both accelerated and focused onto a much smaller fluorescent screen at the top of the vacuum tube. The acceleration and concentration of the electrons resulted in an extremely bright image at the output phosphor. This allowed the use of much lower mA settings for the original x-ray beam, reducing patient dose by two magnitudes (approximately 100 times).

The two most common types of modern fluoroscopes are shown in Figures 14-1 and 14-2. *Dynamic* flat-panel detector (DFPD) units (Figure 14-1) use precisely the same technology as the flat-panel detector (FPD) used for digital radiography discussed in Chapter 10, but with some changes in the sizes of the dexels and the active-matrix array (AMA), and electronics, to accommodate a continuous, moving image. These units send the acquired image directly to display. The older image intensifier (Figure 14-2) is bulkier, but also cheaper, and continues in use. The image intensifier requires a recording camera to be mounted on top of it in order to convert the light

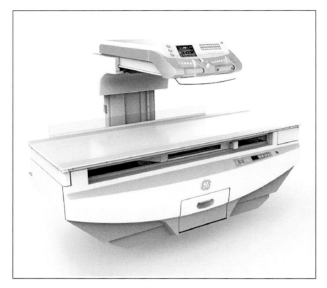

Figure 14-1. Dynamic flat-panel digital (DFPD) fluoroscopy unit.

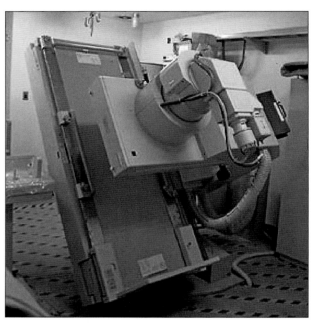

Figure 14-2. Conventional image intensifier fluoroscopy unit.

image emitted from it into an electronic signal to be sent to the display monitor.

Image Recording from an Image Intensifier Tube

Cine film cameras or electronic TV cameras could be mounted on top of the image intensifier to make recordings of the fluoro images. These recording devices have now been largely replaced by a *charge-coupled device (CCD) camera* (Figure 14-3). It consists of a small flat plate of semiconductor material with several prongs for electrical connections along one or two sides. Normally about 1 cm square in size for typical photographic cameras, a larger 2.5 cm version is used to mount atop an image intensifier tube to match its 2.5 cm output phosphor, coupled together by a short bundle of optic fibers.

Figure 14-4 is a diagram of how a typical CCD works. This is very similar to the direct-conversion DR detectors described in Chapter 10. When light photons enter the semiconductor layer, they ionize silicon molecules by the photoelectric effect, ejecting electrons from them and leaving behind positively-charged

Figure 14-3. A charge-coupled (CCD) imaging device. The CMOS looks very similar. (Courtesy, Apogee Instruments, Inc.)

"holes," orbital vacancies in the atoms. A layer of microscopic electrodes beneath the silicon is given a negative electrical charge, and a dielectric layer above is charged positive. This attracts the freed electrons upward. As a sequence of

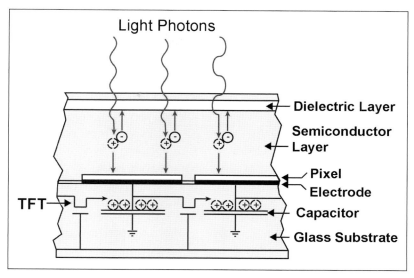

Light Photons

Dielectric Layer

Semiconductor Layer

Pixel

Electrode

Capacitor

Glass Substrate

TFT

Figure 14-4. Components of a charge-coupled device (CCD). When light ionizes the semiconductor layer, electrons drift upward to the dielectric layer, while positively-charged "holes" drift downward to the electrode. Positive charge is stored on a capacitor to form the electronic image signal. (From Q. B. Carroll, *Radiography in the Digital Age,* 3rd ed. Springfield, IL: Charles C Thomas, Publisher, Ltd., 2018. Reprinted by permission.)

electrons flows upward through the semiconductor, the positive vacancies or "holes" in the molecules drift downward. When a "hole" reaches the bottom of the device, it pulls an electron from the electrode there, leaving a positive electric charge on it. Positive charge is collected by capacitors in this bottom layer and stored as an indication of how much x-ray exposure was received.

Underneath the detection layer, rows and columns of thin-film transistors (TFTs) form an active matrix array (AMA). Each TFT acts as an electronic gate. When these gates are opened by a small bias voltage, the electronic charge from each associated capacitor is released in sequence. As this electronic signal is drained from each row of TFTs, it is boosted by an amplifier before passing into the computer as image data.

A typical 2.5 cm CCD coupled to an image intensifier has a matrix of 2048 x 2048 pixels, making each pixel only 14 microns in size and yielding very high sharpness. This amazing level of miniaturization is possible because for the CCD, *each TFT effectively constitutes a pixel,* where-

as for DR each dexel, constitutes a pixel. Rather than individual dexels, each having its own detection surface, the entire active matrix array of TFTs lies underneath a *continuous detection surface.* Compared to the older TV cameras, CCDs also have high detective quantum efficiency (DQE) so that less radiographic technique can be used to save patient dose. Digital fluoro systems must have a signal-to-noise ratio (SNR) of at least 1000:1. CCDs have a high signal-to-noise ratio (SNR), and a long dynamic range of 3000:1, making them ideal for fluoroscopy. CCDs can acquire images at 60 frames per second, twice the frame rate of conventional TV cameras, making CCDs especially suitable for digital fluoroscopy.

Complimentary metal-oxide semiconductors (CMOS's) use two metal-oxide semiconductor field transmitters (*MOSFETs*) stacked together to form an electronic logic gate. One MOSFET is a *p-type* (positive) transistor and the other an *n-type* (negative) transistor. Developed contemporaneously with CCDs in the 1970s, MOSFETs quickly became the preferred method for manufacturing integrated circuits for computers. It was found the CMOS's could be used to collect electric charge liberated from a layer of silicon above just like CCDs, but early CMOS cameras produced inferior images. Recent improvements in CMOS detectors have brought them back into stiff competition with CCDs as camera elements.

Initial capture of the light image for the CMOS works just like that of the CCD. Shown in Figure 14-5, the main difference is that the CCD consists of a single, *continuous* detection surface with an underlying active matrix array made up *only of TFTs*. The CMOS sensor consists of individual dexels each of which has its own separate detection surface. The CMOS is very similar to a DR detector (see Figure 11-4, page 124), but has even more electronics embedded within each dexel including a TFT, an amplifier, a noise correction circuit and a digitization circuit. This results in a much reduced sensitive surface area dedicated to light capture (Figure 14-5, *right*). The lower *fill factor* for CMOS results in higher noise and less uniformity. But, with efficient conversion of light energy into electronic signal, the CMOS has higher speed than the CCD, consumes only about *1/100th* the power, and is much less expensive to manufac-

ture. Note at the *left* in Figure 14-5 that *for the CCD, each TFT in the underlying AMA effectively becomes a pixel.* These individual TFTs are much smaller than the *dexels* of the CMOS (right), resulting in higher spatial resolution for the CCD. The CMOS has higher contrast resolution.

Also shown in Figure 14-5, *readout* of data from the CCD is different than that for the CMOS. In a CCD, each row of TFTs sends a stream of electric charge as an *analog* signal to the corner of the active matrix array. There, the fluctuations in analog signal are separated into digital readings that will represent pixels. In the CMOS sensor, initial output from each dexel has already been separated into a discrete measurement and digitized before moving down the data wire.

Dynamic Flat-Panel Detectors

Flat-panel technology was fully described in Chapter 10 under the headings of *Direct-capture digital radiography* and *indirect-capture DR*. These very same types of detector elements can be arrayed to form a *dynamic* flat-panel detector to record continuous motion images in place of the older image intensifier tube for fluoroscopy. As with DR, *direct-capture* DFPDs use amorphous

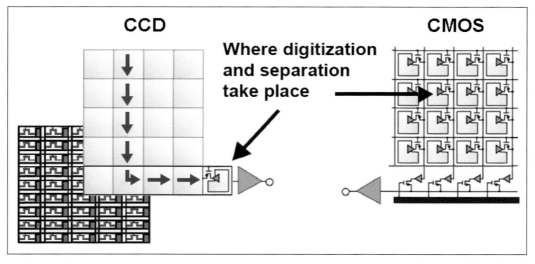

Figure 14-5. The CCD has a *continuous* detection surface with an underlying matrix of small TFTs. The CMOS has lower *fill factor* because each dexel has its own detection surface crowded out by electronics. Separation of the analog signal into digital data occurs at the corner of the entire array for the CCD, but within each inidividual dexel for the CMOS.

Figure 14-6. For an indirect-capture DR system, cesium iodide crystals in the phosphor layer can be formed as long tube-shaped channels to reduce the lateral dispersion of light. (Courtesy, CareStream Health.)

selenium dexels, and *indirect-capture* DFPDs use amorphous silicon dexels positioned under a phosphorescent scintillator layer that first converts x-rays to light. The scintillation layer can be made of cesium iodide or of gadolinium. Cesium iodide can be formed into rod-shaped crystals that keep light from diffusing, as shown in Figure 14-6, but direct the light straight downward to improve resolution. Gadolinium crystals are *turbid,* meaning they cannot be shaped into rods, so gadolinium detectors have slightly less resolution but are cheaper.

Realtime observation of a moving fluoroscopic image does not require the high level of spatial resolution that detailed examination of a static (still) image does. The dexels for a regular DR flat panel are typically 100–150 microns (0.1–0.15 mm) in size. For a dedicated fluoroscope, the dimensions of the dexels are 200–300 microns, twice those used for static DR imaging, resulting in four times the area. *Dual use* digital systems have the 100–150 micron dexels for higher-resolution "spot" images, then bin together groups of four hardware dexels each to form an *effective pixel* with dimensions of 200– 300 microns for the fluoroscopic mode of operation.

The entire dynamic flat panel is usually a large 43 x 43 cm (17 x 17") square, with matrix sizes up to 2048 x 2048 dexels. For an indirect-conversion system, it is the TFT dexel layer, not the scintillation layer, that becomes the limiting

factor for sharpness. The dynamic flat-panel detector sends the image directly to display, with a typical resolution of about 2.5 LP/mm (line-pairs per millimeter). This is only one-half as good as a state-of-the-art image intensifier at 5 LP/mm. However, the image intensifier tube must have a CCD camera mounted at the top which lowers the final resolution of the displayed image down to a level close to that of the DFPD. Depending on the quality of the system, either one may have better sharpness.

To avoid any "ghost image" from a previous frame appearing on a current frame, complete erasure of the DFP detector is necessary after every exposure. This is achieved with a light-emitting diode (LED) array that is located below the detector. Immediately after each frame is sent into the computer system, the LED array produces a bright microsecond flash of light that erases the entire detector in preparation for the next exposure.

Radiographic grids for DFPDs are often constructed with their grid lines running diagonal to help prevent aliasing artifacts (see Chapter 5). Because of the impressive ability of digital processing to compensate for the effects of most scatter radiation, lower ratio grids (6:1 or 8:1) can be used for digital fluoroscopy that will allow reduced technique and decreased patient dose, and the grid can often be removed for imaging children.

Perhaps the greatest advantage of the DFPD over the image intensifier tube is its complete lack of geometrical distortion, whereas the image intensifier suffers from pincushion distortion and vignetting that affect the peripheral portions of the image. The DFPD is smaller, lighter, and easier to use, and requires less power. It has high signal-to-noise ratio (SNR) and extended dynamic range. However, it is more expensive, may have defective dexels, and can have lower spatial resolution.

Reducing Patient Dose

All fluoroscopic procedures should be done using *intermittent fluoroscopy,* where brief visual checks are made from time to time rather than

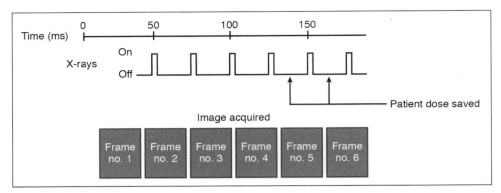

Figure 14-7. By pulsing the fluoroscopic x-ray beam, patient dose is spared from continuous radiation wasted between exposure frames. Using shorter exposure times for the pulses themselves further reduces patient dose. (From Q. B. Carroll, *Radiography in the Digital Age,* 3rd ed. Springfield, IL: Charles C Thomas, Publisher, Ltd., 2018. Reprinted by permission.)

continuously running the fluoroscope, whenever possible without sacrificing the objectives of the study. The 5-minute timer is designed to remind the operator when beam-on time is becoming excessive for GI studies, and should not be overridden by the radiographer. Minimum OID must always be used during fluoroscopy by keeping the IR down as close to the patient as practical. In recent years, there have arisen several cases of actual severe radiation burns and lesions to patients from cardiovascular and surgical c-arm procedures. Radiographers and radiology assistants working in these areas must become familiar with FDA recommendations and exercise all due care to prevent excessive radiation exposure to patients.

An excellent digital tool to assist in minimizing patient exposure is the *last-image hold* feature now available with most digital fluoroscopes. After a short burst of x-rays, a single frame image is displayed continuously until the next exposure is acquired. This feature has been shown to reduce fluoroscopic beam-on time by as much as 50 to 80 percent.

Finally, nearly all digital fluoroscopes now provide an option to be operated in *pulsed fluoroscopy* mode. Although DFPDs can be operated in continuous x-ray mode, this results in an unnecessary and often excessive amount of radiation dose to the patient. Figure 14-7 illustrates fluoroscopic operation in pulsed mode, in which short bursts of radiation are used to produce individual frames. Note that substantial radiation dose to the patient is saved every time the x-ray beam is shut off between frames. To the human eye, as long as 18 frames per second are produced, there is no apparent "flicker" in the continuous image, which appears as a smoothly flowing motion picture. Therefore, any radiation used to produce more than 18 frames per second is wasted.

To achieve very short pulse exposure times on flat-panel C-arm units, mA must be increased as much as 10 times the conventional rate (from 2 mA to 20 mA) in order to obtain sufficient exposure per frame. Even with this increase in mA, the net result is saved patient dose because of the extremely short frame exposure times. High-frequency generators are thus required with very fast interrogation and extinction times. *Interrogation time* is the time required for the x-ray tube to be switched on, and *extinction time* is the time it takes to switch the tube off.

With digital equipment, patient dose can be saved by reducing the pulsed frame rate below 30 frames per second. From the digital image memory, *interpolated* frames can be added in between actual exposed frames to produce a continuous output video signal. This allows us to reduce the actual exposure frame rate to 15 fps, cutting patient dose *in half.* Even frame rates as low as 7.5 fps can be used for some procedures that do not require very high resolution. When in pulse mode, the rate of frames per second is automatically adjusted so as to not exceed the rate of electronic pulses per second.

On some units, further reduction of patient dose can be achieved by reducing the *pulse width* for each frame. This is a capability for digital imaging that needs to be more widely recognized. The *pulse width* is the individual exposure time for each frame exposure, and typically can be set at 6 milliseconds or 3 milliseconds. By selecting a 3 msec pulse width and keeping the mA at 20, patient exposure can again be cut to *one-half*. Newer fluoroscopy units combine high mA pulsed mode with *spectral filters* made of copper or copper/aluminum alloy to achieve increased image quality while keeping patient dose low.

Chapter Review Questions

1. Dynamic flat panel detectors use precisely the same technology as the flat-panel detectors used in _____, with some modifications to accommodate a continuously moving image.

2. The CCD used for fluoroscopy is identical to that used in a digital camera, except for attachment to an image intensifier it must be _____ in size.

3. In a CCD or CMOS, ionization of atoms by x-rays leads to a build-up of electrical _____ collected by capacitors underneath.

4. Underneath the continuous detective surface of a CCD, rows and columns of ____ form an active matrix array (AMA) to process the signal.

5. For the CCD, since each TFT effectively constitutes a pixel, these pixels are only ____ microns in size, making it possible to produce extremely high sharpness.

6. The main difference between a CCD and a CMOS is the way the stored electric charges are ____ off of the chip.

7. In the CMOS, since each dexel includes electronics inside it, CMOS has a much lower _____ *factor* than CCDs, making it less uniform and more noisy.

8. List three practical advantages to the CMOS over the CCD:

9. Cesium iodide DFPDs have _____-shaped crystals that reduce light dispersion, but gadolinium DFPDs are much cheaper.

10. Dual use DFPDs can operate in fluoro mode or in "spot" mode for still images. For efficiency during fluoro mode operation, they *bin together* groups of 4 hardware dexels to form effective pixels that have _____ the dimensions of those used during "spot" mode.

11. Grids designed for use with DFPDs typically have their grid lines running _____ to prevent aliasing.

12. The greatest advantage of the DFPD over the image intensifier tube is its complete lack of geometrical _____.

13. In recent years there have been several cases of severe radiation _____ to patients during cardiovascular and C-arm procedures.

14. On digital fluoro units, the *last-image* _____ feature has been demonstrated to reduce beam-on time as much as 50 percent.

15. The human eye sees no "flicker" at frame rates of at least _____ frames per second.

16. By operating a fluoroscopy unit in _____ mode, unnecessary radiation exposure between frames is saved.

17. The time required for the x-ray tube to be switched on is called the _____ time.

18. On some units, further reduction in patient dose can be achieved by cutting the *pulse* _____ in half.

Chapter 15

QUALITY CONTROL FOR DIGITAL EQUIPMENT

■ ■

Objectives

Upon completion of this chapter, you should be able to:

1. Describe basic quality control tests for a digital x-ray unit.
2. Describe basic quality control tests for electronic image display monitors.

Quality control (QC) is generally understood to be the calibration and monitoring of equipment. It is but one aspect of *quality assurance,* a managerial philosophy that also includes all aspects of patient care, image production and interpretation, workflow, safety, and radiation protection. Calibration of x-ray equipment is largely the domain of the medical physicist. The quality control technologist may be trained in most of these procedures. Even though the average staff radiographer may never perform most of these equipment tests, it is important that the radiographer understand and appreciate the required types of tests and why some acceptance parameters are more stringent than others. Every staff radiographer should have a level of understanding that enables him or her to at least recognize problems that may be related to QC and report them to the proper supervisor, physicist, or QC technologist.

Monitoring of Digital X-Ray Units

Some tests are easy enough for the staff radiographer to perform. Once a baseline image or data set is established for a particular x-ray unit or display monitor, it can be monitored by technologists for any sudden or dramatic deviation by regular, simple visual checks.

Field Uniformity

All digital x-ray detectors are inherently nonuniform. In Chapter 5 we discussed how digital preprocessing is used to correct for deviations in intensity across the field, and for dexel drop-out in DR detectors. Any time a particular radiopaque (light) artifact is noted repeatedly in the same spot on DR images from a particular unit, the DR detector plate used in the room is suspect. Similar CR artifacts can be attributed to a specific cassette when they repeatedly appear on images produced with that cassette, or to the CR reader if they repeatedly appear on images taken with various cassettes. As explained in Chapter 5, noise reduction algorithms correct for these types of flaws detected when a digital x-ray unit is initially installed, but new artifacts will eventually appear due to "wear and tear" of imaging plates and equipment as they age. These should be monitored and reported.

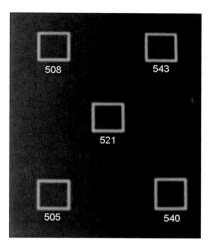

Figure 15-1. Field uniformity is measured by comparing the center and four corners of a "flat-field" image, an x-ray exposure with no object in the field.

Field uniformity from the image receptor can be tested by making a "flat field" exposure with no object in the x-ray beam, using moderate technique settings. The plate should be thoroughly erased prior to any test exposure (DR systems automatically erase the detector after each exposure). A long SID of 180 cm (72″) should be used to minimize the anode heel effect. The resulting image may be visually scanned for dexel or pixel defects. Uniform field intensity is measured at five locations across the large (35 x 42 cm) field—the center, and four corner areas as illustrated in Figure 15-1. All images have a degree of non-uniformity which will become apparent at extreme contrast settings, so contrast must not be arbitrarily increased by reducing the window width when evaluating these test images.

Intrinisic (Dark) Noise

Intrinsic or *dark noise* is the statistical noise inherent to the DR detector that is not related to individual dexel drop-out, or "background" noise inherent in the CR phosphor plate which is not related to the reader. For CR, a single phosphor plate can be erased and processed without any exposure to x-rays. Visually scan the image for unusual amounts of mottle or noise.

Erasure Thoroughness and "Ghosting"

In a CR reader, the erasing chamber uses bright bulbs to expose the phosphor plate and purge it from any residual image from the last exposure. It is possible for the erasure time to be too short, leaving some residual image. The light bulbs can also lose intensity over time, or burn out. Any of these problems would result in insufficient erasure of the plate. Note that even with proper erasure, when a CR phosphor plate is subjected to an extreme x-ray exposure, a residual image may persist.

To check for inadequate erasure, first expose an aluminum step wedge or other object made of homogeneous material and process the PSP plate, then immediately make a second exposure on the same plate without the object in the beam and with about 2 cm of collimation on all sides of the field. The image can then be visually examined for any appearance of the object as a "ghost image" or "image lag."

Ghosting is also known as "memory effect" in DR systems, and can be checked the same way. In a DR detector, ghosting is caused from residual electrical charge that has been trapped in the amorphous selenium or silicon detection layer, and is released only slowly over time. Indirect-conversion DR systems usually have shorter image lag than direct-conversion systems.

Sharpness

A fine wire mesh can be exposed and the image visually examined for any blurry areas within the field. A baseline image should be made when new equipment is installed, then regular test images taken every 6 months can be compared to this standard. A line-pair test template (see Figure 3-10 in Chapter 3) can be used to measure the resolved spatial frequency in line-pairs per millimeter (LP/mm) for any system.

Monitoring of Electronic Image Display Systems

In the digital age, the electronic display monitor is typically the weakest link in the entire imaging chain. Poor spatial resolution or contrast resolution can lead to misdiagnosis, so quality control of the display monitor has become extremely important. Every display monitor in an imaging department should be quality-checked at least once each year. Monitors should be cleaned each month with very mild detergent and a soft, non-abrasive cloth. Consistency between all display monitors within a department is essential. Especially within a radiologist's workstation, the two high-resolution monitors must have precisely equal luminance, contrast, and resolution characteristics for diagnostic use.

QC standards and guidelines for display devices are available from several scientific groups. Specific standards for medical images are found in DICOM Part 14. The American Association of Physicists in Medicine (AAPM), the American College of Radiology (ACR), and the So-

ciety of Motion Picture and Television Engineers (SMPTE) have published test procedures and guidelines. *Class 1* workstation monitors used by radiologists for diagnosis are subject to more stringent guidelines than *class 2* display monitors used by technologists, other physicians and other departments. Task group 18 of the AAPM and others provide test patterns that can be downloaded from their websites for display on a particular monitor. One of the most comprehensive and useful test patterns was developed many years ago by the SMPTE and is demonstrated in Figure 15-2.

The concepts of *luminance, illuminance, specular reflectance, diffuse reflectance,* and *ambient lighting* for the medical image reading room were introduced and defined in Chapter 11, with recommended guidelines. Following are the most common specific test procedures and the acceptable limits of deviation associated with them.

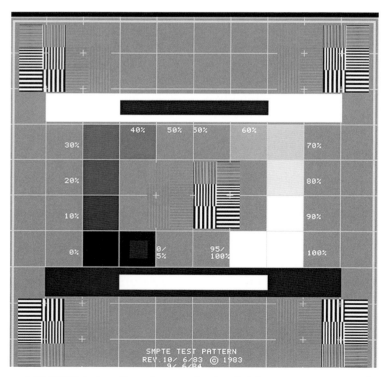

Figure 15-2. The universal SMPTE test pattern for monitoring brightness, contrast, and resolution of electronic image display monitors. (Courtesy, Society of Motion Picture and Television Engineers.)

Maximum Luminance

The *maximum luminance* (maximum brightness) of a display monitor is simply measured with a photometer (Figure 12-17 in Chapter 12) over the brightest area of a test pattern. The American College of Radiology requires a minimum brightness of 250 lumens (Lm) for display systems. But radiologists prefer settings from 500 to 600 Lm, and most LCD monitors can achieve up to 800 Lm. Therefore, we recommend here an acceptable maximum luminance of at least 600 Lm.

Luminance Response

The *luminance response* is an accuracy test that checks the monitor's ability to reproduce different shades of brightness from a test pattern. Test patterns typically consist of adjacent squares with differing shades of gray just above the threshold of human vision to detect. On a monitor with good luminance response, these squares will be resolved with a visible difference in density. On a monitor with poor response, two adjacent squares will appear as one rectangle presenting a single density.

DICOM standards define a *grayscale standard display function (GSDF)* that brings digital values into linear alignment with human visual perception. The increments for the GSDF are called *JNDs,* for "just-noticeable differences." The AAPM recommends that luminance response should fall within 10 percent of the GSDF standard.

The central portion of the SMPTE pattern shown in Figure 15-2 includes a ring of density squares (similar to the image of a step wedge) forming a "ramp" with 10 percent increments from black to white. These should all be within 10 percent accuracy, based on starting measurements of the 0 percent and 100 percent areas using a photometer. The two "50 percent" squares at the top of the ring should match. At the bottom of the ring there are two gray squares in the middle used only for labeling the double-squares lateral to them, which consist of a smaller square *within* a larger square. The double-square to the left is dark gray within black on the left, marked "0/5 percent." The one on the right is light gray within white, marked "95/100 percent." The observer should be able to distinguish the smaller square within the larger one. In this particular image, the monitor "passes" for the dark squares at the left, but seems to fail for the light squares on the right, indicating somewhat excessive brightness.

Luminance Ratio

Luminance ratio compares the maximum luminance (L_{MAX}) to the minimum luminance (L_{MIN}) and is essentially a contrast check. LCD (liquid crystal display) monitors are very poor at producing a "true black" level (L_{MIN}) while energized, (much worse than the older CRT television monitors), and this tends to result in poor overall contrast. The typical luminance ratio for an LCD falls between 300 and 600. The initial LR of a monitor should be logged, and then checked over time for gradual deterioration. On the SMPTE test pattern in Figure 15-2, two large horizontal double-bars are provided specifically to measure luminance ratio, a black-on-white bar above the center area, and a white-on-black bar beneath.

Luminance Uniformity

Luminance uniformity refers to the consistency of a single brightness level displayed over different areas of the screen. As was done with flat-field uniformity of the detector, luminance uniformity of a monitor is checked at five locations: the center of the screen and the four corners. AAPM guidelines state that these should not deviate by more than 30 percent from their average.

Reflectance and Ambient Lighting

Diffuse and *specular* reflectance off a monitor screen were described in Chapter 12. The ambient lighting in a medical image reading room must be dimmed to a point where both types of reflectance are below any visually noticeable level. While diagnosis is taking place, the maximum ambient lighting in a reading room should never be more than 25 lux. (The *lux* unit is defined as 1 lumen per square meter.)

This is approximately one-fourth the typical brightness of normal office lighting (75-100 lux). Although the illuminance of a reading room is most often adjusted subjectively, specific guidelines and scientific methods for establishing the best level for diagnostic purposes in an objective fashion are available.

Noise

Figure 15-3 demonstrates how excessive noise can affect the visibility of subtle images by its destructive impact upon contrast. This test pattern from the AAPM presents five dots (a central dot and one in each corner) with one *JND*, or *just-noticeable difference* against the background as defined by the DICOM grayscale standard display function (GSDF). On the left in Figure 15-3, one can identify at least two of these dots. In the high-noise image at the right none of the dots are visibly resolved.

Objective tests and measurements have been developed by physicists for the noise level in an image. However, *no quantified guidelines for acceptable levels of noise for medical images have been published,* and with good reason: Recall from Chapter 9 that in the interest of minimizing patient dose, we allow for a certain amount of "acceptable mottle" in the image as defined by the radi-

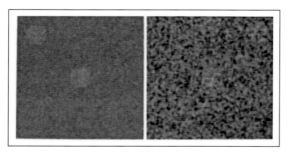

Figure 15-3. A QC test pattern with five dots at one *JND (just-noticeable difference)* evaluates the level of noise in an electronically displayed image. Here, only two of the five dots can be made out in the left image. Excessive noise levels (*right*) can destroy visibility by reducing contrast. (Courtesy, American Association of Physicists in Medicine.)

ologist. Although noise in general should be kept as low as possible, some forms of noise are inevitable and become a secondary consideration in comparison with more important objectives in patient care.

Resolution

Examining the universal SMPTE test pattern in Figure 15-2 one final time, we note in each corner and at the center a series of high-contrast lines or bars laid out in two columns, one for vertical bars and one for horizontal bars, and in three sets of diminishing width. A closer look to the side of these series will reveal another identical set of line patterns at extremely low contrast. These are for testing *sharpness* at these five points of the display monitor screen. Vertical bars indicate *horizontal* resolution, while the horizontal bars are used to check *vertical* resolution which can differ. The observer inspects the sets of lines to determine which is the narrowest set of lines that can be clearly distinguished *before* they become blurred into each other and lose their apparent separation. A table is then consulted to find the LP/mm measurement associated with that set of lines.

The resolution of an LCD monitor is consistent over time and need not be repeatedly checked, but these measurements can be of great value when initially purchasing display monitors. For electronically displayed images, both the AAPM and the ACR recommend a minimum resolution of 2.5 line-pairs per millimeter (2.5 LP/mm).

Dead and Stuck Pixels

In Chapter 11 we described how, for liquid crystal display (LCD) monitors, electrical charge is applied to turn a hardware pixel "off" so that it is dark against a lighter background. Truly "dead" pixels are failing to receive electric current, and actually show up as *white* spots. To test for them, a *black* background "flat field" image must be created, which would be the normal image from an x-ray exposure with no object placed in the beam. It is also possible for a hardware pixel to become *stuck* in its "off" state. In this case, it is receiving continuous electrical current and shows up as a black spot against a light background. Such a background is the default mode for most word processing software run on your desktop computer, presented as a "blank page." For medical imaging monitors, a blank white screen specifically for test purposes may be available in a special file.

Figure 12-18 (Chapter 12) is a diagram illustrating how a hardware display pixel is actually made up of three smaller *subpixels*. An entire hardware pixel is typically the size of a printed period or the dot of an "i" in 12-point font. With good vision, a bad pixel is readily apparent upon close inspection of the monitor screen. A magnifying glass will help with initially finding bad pixels, but having located one, it can then be made out without magnification. A single subpixel can also fail, but would only be visible with the aid of magnification.

Dead or stuck pixels can sometimes be fixed by very gently massaging them with the fleshy fingertip. However, since a monitor pixel can also be damaged by fingernails or by excessive pressure, this procedure is not recommended for expensive workstation monitors.

It is not unusual for any LCD monitor to have a few bad pixels present. For *Class 1* monitors used by radiologists for diagnosis, manufacturers recommend that the LCD be "nearly perfect." For *class 2* monitors, they recommend a threshold of 2 defective pixels, or 5 defective subpixels, per million. The AAPM recommends failing QC for an LCD monitor that has more than 15 bad pixels overall, more than 3 bad pixels within any 1 cm circle, or more than 3 bad pixels adjacent to each other anywhere on the screen.

Other QC Tests

Viewing angle dependence has been mentioned as a major disadvantage for LCD monitors. Objective measurements can be made by a physicist using a photometer. The *stability* of a new self-calibrating LCD monitor should be checked by measuring the maximum brightness

every day for the first week, once a week for a month, then once a month for three months. Luminance tests should be performed on all medical monitors annually.

Note, too, that there are several "non-digital" QC tests that should be performed on x-ray machines at least once each year, and fall outside the scope of this textbook. They are described in most x-ray physics books.

Chapter Review Questions

1. For CR units, an artifact that appears in the same location when different cassettes are used is most likely being caused in the _____.

2. For a flat-field uniformity test, the brightness or density of the image is measured in _____ locations across the field for comparison.

3. General statistical noise that is inherent to a DR detector is called _____ noise.

4. Two exposures quickly taken from the same detector, one with an object in it and one without, are used to test for _____.

5. Exposures of a fine wire mesh can be used to periodically check for any changes in _____ across the area of a detector.

6. Display monitors should be quality-checked at least once each _____.

7. The luminance of a display monitor can be measured using a _____.

8. For a display monitor, list four different quality control tests related to *luminance*.

9. The QC test that checks a display monitor's ability to reproduce different shades of brightness is called *luminance* _____.

10. For a display monitor, the L$_{MAX}$ divided by the L$_{MIN}$ is the *luminance* _____.

11. On a display monitor, a pixel that appears dark against a light background is _____.

12. For display monitor QC, there should never be _____ bad pixels within a 1 cm circle, or _____ bad pixels adjacent to each other.

GLOSSARY OF DIGITAL RADIOGRAPHIC TERMS

Absorption Efficiency: The ratio of photons absorbed by a particular material.

Accession Number: An institution's unique number for identifying a particular procedure or exam.

Active Matrix Array: A panel of electronic detector elements laid out in rows and columns, used to convert incoming light or x-ray photons into an electrical signal.

Algorithm: A set of computer instructions to execute a specific task. In radiography, the "procedural algorithm" often refers to the default processing codes for each particular anatomical part.

Aliasing: See *Moire effect.*

Artifacts: Any extraneous images that obscure the desired information in a radiograph.

Attenuation: Reduction in the number or intensity of x-rays as a result of absorption and scattering.

Attenuation Coefficient: The ratio or percentage of the original x-ray beam intensity that is absorbed by a particular tissue area.

Band-pass filtering: Removal of selected frequency layers when a digital image is reconstructed by inverse Fourier transformation.

Binary: Having only two values, 0 and 1, yes or no, on or off.

Bit: *Binary digit,* a zero or a 1, the smallest unit of computer storage capacity.

Bit Depth: The maximum range of pixel values a computer, display monitor, or other digital device can store.

Byte: A group of 8 bits used to represent a single alphanumeric character.

Cassette: Any rigid, flat rectangular container designed to hold a PSP plate or film.

Central Processing Unit (CPU): The computationally functional hardware of a computer system, including the control unit, arithmetic-logic unit, and primary storage.

Characteristic Curve: A graph of the optical density produced in a displayed image plotted against the exposure level that produced the original latent image.

Charge-Coupled Device (CCD): A flat, compact light-sensing device that uses a single layer of silicon as its sensitive surface, used for recording images.

Complimentary Metal-Oxide Semiconductor (CMOS). A flat, compact light-sensing device that uses two semiconductor layers stacked together, one positive and one negative, used to record images.

Computed Radiography (CR): Use of a photostimulable phosphor plate as an image receptor, which, after exposure, is scanned by a laser beam to release a latent light image. This light image is then converted into an electronic signal that can be fed into a computer for processing manipulation.

Computed Tomography (CT): Reconstruction of a cross-sectional image of the body using multiple projections from a fan-shaped x-ray beam.

Contrast: The ratio or percentage difference between two adjacent brightness (density) levels.

Contrast Resolution: The ability of an imaging system to distinguish and display a range of attenuation coefficients from different tissues within the body; the ability to reproduce subject contrast with fidelity.

Conversion Efficiency: The percentage of energy from absorbed x-ray photons that is converted to light by a particular phosphor material.

Current, Electrical: Intensity of the flow of electrons.

Cycle, Wave: Smallest component of a waveform in which no condition is repeated, consisting of a positive pulse and a negative pulse.

Dark Masking: Surrounding the image with a black border to enhance perceived contrast.

Del: See *Dexel*

Density, Electron: The concentration of electrons within a particular atom.

Density, Physical: The concentration of mass (atoms) within an object.

Density, Radiographic: The degree of darkness for an area in the image.

Detail: The smallest component of a visible image.

Detective Quantum Efficiency (DQE): The ability of a detector element (del) to absorb x-rays or light; its sensitivity to photons.

Deviation Index: A standardized read-out indicating how far an exposure falls outside the target or "ideal" exposure level expected for a particular projection.

Dexel: Acronym for "detector element," an individual hardware cell in a DR image receptor, capable of producing a single electronic readout from incoming photon (light or x-ray) energy.

DICOM: Digital Imaging and Communications in Medicine guidelines which standardize the behavior of all the various digital devices used in PAC systems.

DICOM Header: A series of lines of information from the metadata of an image that can be displayed at the touch of a button at the monitor, or may be routinely displayed at the top (head) of each image.

Differential Absorption: Varying degrees of x-ray attenuation by different tissues that produces subject contrast in the remnant x-ray beam.

Digital Fluoroscopy (DF): Real-time (immediate), dynamic (motion) imaging with an area x-ray beam and image intensifier, in which the final electronic signal is digitized for computer processing.

Direct-Capture Digital Radiography (DR): Use of an active matrix array (AMA) as an image receptor, which converts x-ray exposure into an electronic signal that can be fed into a computer for processing manipulation. A DR system does not require the exposed image receptor (cassette) to be physically carried to a separate processing unit.

Direct-Conversion DR: A DR unit whose image receptor converts incoming x-rays directly into electrical charge with no intermediate steps.

Display Station: A computer terminal restricted to accessing and displaying radiologic images.

Distortion, Shape: The difference between the length/width ratio of an image and the length/width ratio of the real object it represents, consisting of elongation or foreshortening.

Distortion, Size: See *Magnification.*

Dot Pitch: See *Pixel Pitch.*

Dual Energy Subtraction: The production of "bone-only" or "soft-tissue only" images by using two exposures at different average energies (average keV). This allows the computer to identify tissue types by the change in x-ray attenuation, then subtract bony or soft tissue from the displayed image.

Dynamic: Moving.

Dynamic Range: The range of pixel values made available for image formation by the software and hardware of an imaging system.

Dynamic Range Compression: Removal (truncation) of the darkest and lightest pixel values from the gray scale of a digital image.

Edge Enhancement: Use of spatial or frequency methods to make small details, such as the edges of structures, more visible in the image.

Emission Efficiency: The percentage of light produced by phosphor crystals that escapes the phosphor layer and reaches a detector.

Energy Subtraction: See *Dual energy subtraction.*

Equalization: See *Dynamic range compression.*

Exposure Field Recognition: Ability of a digital imaging system to identify the borders of a collimated x-ray exposure field, so that data outside the field may be excluded from histogram analysis and exposure indicator calculations.

Exposure Indicator: A read-out estimating the exposure level received at the image detector as derived from initial pixel values in the acquired image histogram.

Exposure Latitude: The margin for error in setting radiographic technique that an imaging system will allow and be able to produce a diagnostic image.

Extrapolation: Estimation of a value beyond the range of known values.

Farbzentren (F Centers): Energy levels within the atoms of a phosphor crystal that act as electron "traps" that can store an ionized electron for a period of time. Collectively, these captured electrons form a latent image.

Fast Scan Direction: The direction the laser moves across a PSP plate in a CR reader.

Fifteen-Percent Rule: The exposure level at the image receptor can be approximately maintained while increasing x-ray beam penetration, by employing a 15 percent increase in kVp with a halving of the mAs, and vice versa.

Fill Factor: The percentage of a detector element's area dedicated to photon absorption.

Flat Field Uniformity: The consistency of pixel levels across the area of the image field when exposure is made with no object or anatomy in the x-ray beam.

Fluoroscopy: Real-time (immediate) viewing of dynamic (moving) radiographic images.

Fog: An area of the image with an excessive loss of contrast and gray scale due to sources of noise.

Fourier Transformation: The mathematical process, in the frequency domain, that breaks the complex waves that form an image into their individual wave components.

Frequency: Rate of cycles per second for a moving waveform.

Frequency Domain Processing: Digital processing operations based upon the size of structures or objects in the image, which correlates to the wavelength or frequency of their wave function.

Geometrical Integrity: See *Recognizability.*

Gradation (Gradient) Processing: Re-mapping of the gray scale range of a digital image by the use of look-up tables (LUTs). Intensity transformation formulas are applied to the range of pixel values to generate new LUTs.

Gray Scale: The range of pixel values, brightness levels, or densities present in a displayed digital image.

Hardware: Physical components of a computer or related device.

Hertz: Unit of frequency; number of cycles or oscillations per second.

High-pass filtering: Removal of low-frequency layers when the digital image is reconstructed, such that higher frequencies (smaller details) become more visible.

HIS: Hospital Information System which stores and manages all images, lab results, reports, billing and demographic information on every patient.

Histogram: A graph plotting the pixel count for each brightness level (density) within an entire image.

Hybrid Subtraction: A combination of temporal and energy subtraction, where the two images are taken at different times and also at different energies, precluding the need of a contrast agent.

Image Receptor: Device that detects the remnant x-ray beam and records a latent image to be later processed into a visible image.

Indirect-Conversion DR: A DR unit whose image receptor first converts incoming x-rays directly into light using a phosphor layer, then converts the light into electrical charge.

Intensity Domain Processing: Digital processing operations based upon pixel values (brightness , or density levels), which form a histogram of the image.

Interface: Connection between computer systems or related devices.

Intermittent Fluoroscopy: Avoidance of continuous exposure to the patient during fluoroscopic procedures.

Interpolation: Estimation of a value between two known values.

Interrogation Time: Time required for an electronic switch or detector to operate.

Inverse Fourier Transformation: The mathematical process, in the frequency domain, that re-assembles or adds together the individual wavelengths

Inverse Square Law: The intensity of radiation (and most forces) is inversely proportional to the square of the distance between the emitter and the receiver.

Inversion, Image: "Flipping" an image top for bottom and bottom for top.

Jukebox: A stack of dozens of optical disc or tape storage devices.

K-Edge Effect: The surge of photoelectric light emission by a phosphor when the kV of incoming photons just exceeds the binding energy of the phosphor atoms K shell.

Kernel: A smaller sub-matrix that is passed over the image matrix executing mathematical operations on its pixels.

Kilovolt-Peak (kVp): The maximum energy achieved by x-ray photons or electrons during an exposure.

Laser: *Light amplification by stimulated emission of radiation:* Light emitted in phase (with all waves synchronized) when certain materials are electrically stimulated.

Latent Image: Image information that has not yet been processed into a visible form.

LCD: See *Liquid crystal display.*

Light Guide: A fiber optic tube that can transmit light along a curved path.

Liquid Crystal Display (LCD): A display monitor that uses a layer of liquid crystals between two polarized sheets of glass to control the blocking or emission of light from behind the screen.

Look-Up Table (LUT): A simple table of data that converts input gray level values to desired output values for the displayed image.

Lossless Compression: Digital image compression ratios less than 8:1 that preserve a diagnostic quality of image resolution.

Lossy Compression: Digital image compression ratios at 8:1 or greater that do not preserve a diagnostic quality of image resolution.

Low-Contrast Resolution: Ability to demonstrate adjacent objects with similar brightness (density) values).

Low-pass filtering: Removal of high-frequency layers when the digital image is reconstructed, such that lower-frequencies (larger structures and backgrounds) become more visible.

Magnification: The difference between the size of an image and the size of the real object it represents.

mAs: See *Milliampere-seconds.*

Matrix: The collective rows and columns of pixels or dels that make up the area of an image or image receptor.

Metadata: An extensive database of information stored for each image in a PAC system, including information on the patient, the institution, the procedure, the exposure and the equipment used.

Milliamaperage: The standard unit measuring the rate of flow of electricity (also frequently used by radiographers to describe the rate of flow of x-ray photons).

Milliampere-Seconds (mAs): The total amount of electricity used during an exposure (also frequently used by radiographers to describe the total amount of x-ray exposure used).

Misregistration: Misalignment of two images for the purpose of subtraction.

Modem: *Modulator/Demodulator:* A device that converts digital electronic signals into analog musical tones and vice versa.

Modulation: Changing of a video or audio signal.

Modulation Transfer Function (MTF): Measurement of the ability of an imaging system to convert alternating signals in the remnant x-ray beam into alternating densities or pixel values.

Moire Effect: False linear patterns produced in an image when the sampling rate of a processor/reader approximates the line resolution of the image receptor or the line frequency of a grid.

Mottle: A grainy or speckled appearance to the image caused by excessive noise, faulty image receptors or faulty display.

Multiscale Processing: Decomposition of the original image, by Fourier transformation, into eight or more frequency bands for individual digital processing treatments.

Nematic Crystals: Crystals with a long thread-like shape, that tend to align together.

Noise: Any form of non-useful input to the image which interferes with the visibility of anatomy or pathology of interest.

Operating System: Software that sets the general format for the use of a computer (e.g., home, business, science) including an appropriate interface ("desktop") for the type of input devices to be predominantly used.

Optical Disk: A removable plate that uses laser light to write and read data.

Overexposure: An area of the image excessively darkened due to extreme exposure or processing conditions.

PACS: Picture Archiving and Communication System which provides digital storage, retrieval, manipulation and transmission of radiographic images.

Penetration: The ratio or percentage of x-rays transmitted through the patient to the image receptor.

Penumbra: Blur; The amount of spread for the edge of an image detail, such that it gradually transitions from the detail brightness to the background brightness.

Perceptual Tone Scaling: Gradation processing with LUTs based on the human perception of brightness levels in order to give the displayed image a more conventional appearance.

Phosphor: A crystalline material that emits light when exposed to x-rays.

Phosphorescence: Delayed emission of light.

Photoconductor: Material that conducts electricity when illuminated by light or x-ray photons.

Photodiode: Solid-state electronic device that converts light or x-ray energy into electrical current.

Photoelectric Interaction: Total absorption of the energy of an incoming photon by an atom resulting in the emission of a only a photoelectron.

Photometer: Device that measures light intensity.

Photomultiplier Tube: A series of dynodes (reversible electrodes) that can magnify a pulse of electricity by alternating negative and positive charges in an accelerating sequence.

Photon: Electromagnetic energy in a discrete or quantized amount that acts as a "bundle" or "packet" of energy much like a particle.

Photospot Camera: Device that records only one frame at a time from a fluoroscopic image.

Photostimulable Phosphor Plate (PSP): The active image receptor used in computed radiograph (CR) whose phosphor layer can be induced to emit its stored light image, after initial exposure to x-rays, by re-stimulating it with a laser beam.

Phototimer: See *Automatic exposure control.*

Pixel: An individual two-dimensional picture element or cell in a displayed image, capable of producing the entire range of gray levels or colors (bit depth) for the system.

Pixel Pitch: The distance from the center of a pixel to the center of an adjacent pixel.

Polarization, Light: The filtering of light waves such that their electrical component is only allowed through in one orientation, vertical or horizontal, but not both.

Preprocessing: All corrections made to the "raw" digital image due to physical flaws in image acquisition, designed to "normalize" the image or make it appear like a conventional radiograph.

Point Processing Operations: Digital processing operations that are executed pixel by individual pixel.

Postprocessing: Refinements to the digital image made after preprocessing corrections are completed, targeted at specific anatomy, particular pathological conditions, or viewer preferences.

Prefetching: An automatic search for and access of previous radiographic studies and records related to a scheduled exam many hours prior to the exam, so that they may be immediately available to the diagnostician at the time of the exam.

Preprogrammed Technique: Pre-set techniques for each anatomical part stored within the computer terminal of a radiographic unit, selected by anatomical diagrams at the control console.

Primary Memory: Data storage that is necessary for a computer to function generally.

Primary X-Rays: X-rays originating from the x-ray tube that have not interacted with any subsequent object.

Proportional Anatomy Approach to Technique: The thesis that the required overall radiographic technique is proportional to the thickness of the anatomical part, such that comparisons between different body parts can be made to derive ratios for adjusting technique.

Pulse Mode Fluoroscopy: Breaking of the fluoroscopic exposure into a series of discrete, fractional exposures separated by time intervals.

Pyramidal Decomposition: Repeated splitting of the digital image into high-frequency and low-frequency components, with the lower-frequency bands requiring smaller and smaller file sizes for storage.

Quality Assurance: Overall program directed at good patient care, customer satisfaction, and maintenance of resources such as equipment.

Quality Control: Regular testing and monitoring of equipment to assure consistent quality images.

Quantization: Assigning digital values to each measurement from the pixels of an image matrix.

Quantum: See *Photon*.

Quantum Mottle: A grainy or specked appearance to the image caused by the randomness of the distribution of x-rays within the beam, which becomes more apparent with low exposure levels to the image receptor.

Radiography: Use of x-rays to produce single-projection images.

Radiolucent: Allowing x-rays to easily pass through.

Radiopaque: Absorbent to an x-ray beam, such that few pass through.

Random Access Memory (RAM): Data that can be accessed from anywhere within storage in approximately equal amounts of time, without having to pass through other files or tracks.

Read-Only Memory (ROM): Data storage that cannot be changed by the user.

Real-time: Immediately accessible.

Recognizability: The ability to identify what an image is, which depends upon its levels of sharpness, magnification and distortion.

Redundant Array of Independent Discs (RAID): A system that stores back-up image files across several computer hard drive that are independent of each other, to protect against loss of files.

Region of Interest (ROI): See *Volume (values) of interest.*

Remnant Radiation: Radiation having passed through the body but not having struck the image receptor or detector.

Rescaling: Re-mapping of the input data from the acquired image to a pre-set range of pixel values in order to produce consistent brightness and gray scale in the displayed output image.

Resolution: The ability to distinguish small adjacent details in the image as separate and distinct from each other and from any background.

Resolving Time: Time required for a radiation detection device to re-set itself after an ionizing event such that a subsequent ionization can be detected.

Reversal, Image: Changing a (negative) radiographic image into a positive image by inverting pixel values.

RIS: Radiology Information System which stores and manages all radiologic images and related reports, radiologic billing and demographic information on every patient.

Sampling: Taking intensity measurements from each pixel or cell in an image matrix.

Saturation, Digital Radiography: Overwhelming a digital data-analysis system with electronic signal (from extreme exposure at the detector) such that the data cannot be properly analyzed, resulting in a diagnostically useless pitch black area within the displayed image.

Scanning: Dividing of the field of the image into an array of cells or pixels, to format the matrix.

Scatter X-Rays: Secondary x-rays that are traveling in a different direction than the original x-ray beam.

Scintillation: Emission of light upon x-ray or electrical stimulation.

Secondary Memory: Data storage that is not necessary for a computer to function generally.

Secondary X-Rays: X-rays produced by interactions within the patient or objects exposed by the x-ray beam.

Segmentation: Ability of a digital imaging system to identify and count multiple exposure fields within the area of the image receptor.

Semiconductor: Material that can act as an electrical conductor or resistor according to variable conditions such as temperature, the presence of light, or preexisting charge.

Service Class Provider: The centralized storage and distribution services of a PAC system.

Service Class User: The decentralized image acquisition units, workstations and display stations of a PAC system where the services of the system are accessed.

Shape Distortion: See *Distortion, shape.*

Sharpness: The abruptness with which the edges of an image detail stop as one scans across the area of an image.

Signal-to-Noise Ratio (SNR): The percentage (ratio) of useful signal or information to non-useful, destructive noise contained within the acquired image data.

Slow Scan (Subscan) Direction: The direction a PSP plate moves through a CR reader.

Smoothing: Use of spatial or frequency methods to reduce contrast only at the level of small details, (local contrast), to reduce noise or make the edges of structures appear less harsh in the displayed image.

Software: Mathematical and logic programming that directs a computer or related devices.

Solid-State: Using crystalline materials rather than vacuum tubes to control electrical current.

Spatial Domain Processing: Digital processing operations based upon the locations of structures or pixels within the image matrix.

Spatial Frequency: Measure of the spatial frequency of an imaging system expressed in line-pairs per millimeter (or per other unit of distance).

Spatial Resolution: The ability of an imaging system to distinguish and display small pixels; the ability to produce *sharpness* in image details.

Speed: The physical sensitivity of an image receptor system to x-ray exposure.

Speed Class: The programmed sensitivity of a digital processor or reader to the signal from the image receptor.

Square Law: Principle that exposure at the IR can be maintained for different SIDs by changing the mAs by the square of the distance change.

Static: Still or stationary, not moving.

Stimulated Phosphorescence: Delayed emission of light upon stimulation.

Stitching, Image: Attaching several images together to display a large portion of the body.

Storage Area Network (SAN): A small local network of storage devices connected in parallel so that if one device fails, all connections between other devices are preserved.

Subject Contrast: Difference of intensity between portions of the remnant x-ray beam, determined by the various absorption characteristics of tissues within the patient's body.

Subpixel: A segment of a pixel that is not capable of producing the entire range of gray levels or colors (bit depth) for the system.

Subtraction: Removal of structures of a particular size from an image by making a positive (reversed image) mask of only those structures, then superimposing the positive over the original negative image to cancel those structures out. A contrast agent present in only one of the images will not be subtracted.

Technique, Radiographic: The electronic, time and distance factors employed for a specific projection.

Teleradiology: Transmission of medical images and reports to and from remote sites.

Temporal Subtraction: Subtraction between two images taken at different times, the first prior to injection of a contrast agent and the second after the contrast agent is introduced.

Thin-Film Transistor (TFT): The electronic switching gate used in detector elements for direct-capture radiography.

Translation, Image: "Flipping" an image geometrically left for right, and right for left.

Translucent: Allowing light to pass through.

Umbra: The clearly defined portion of a projected image.

Unsharp Mask Filtering: Subtraction of the gross (larger) structures in an image to make small details more visible.

Unsharpness: The amount of blur or penumbra present at the edges of an image detail.

Video Display Terminal (VDT): Display monitor and connected input devices for a computer.

Volatile Memory: Data storage that can be erased by the user.

Volume (Values) of Interest (VOI): Portion of an image histogram that contains data useful for digital processing and for calculating an accurate exposure indicator.

Visibility: The ability to see a structure in an image, which depends upon its brightness and contrast versus any noise present.

Voxel: A three-dimensional cube (or square tube) of body tissue that is sampled by an imaging machine to build up a displayed digital image.

Wavelength: Distance between two similar points in a waveform.

Window Level (Level): The digital control for the overall or average brightness, density or pixel level of the displayed image.

Window Width (Window): The digital control for the range of brightness, density, or pixel levels in the displayed image.

Windowing: Adjusting the window level or window width.

Word, Digital: Two bytes of information.

Workstation: A computer terminal allowing access to and manipulation of radiologic images, and permanent saving of changes into the PAC system.

REFERENCES

American Association of Physicists in Medicine, *An Exposure Indicator for Digital Radiography.* Report of AAPM Task Group #116, July, 2009.

American Association of Physicists in Medicine, *Assessment of Display Performance for Medical Imaging Systems.* Report of AAPM Task Group #18, July, 2010. www .deckard.mc.duke.edu/~samei/tg18

American Association of Physicists in Medicine, Report No 90, *Partitioned Pattern and Exposure Field Recognition.* www.aapm.org/pubs/reports/RPT_93.pdf

American Association of Physicists in Medicine, *Recommended Exposure Indicator for Digital Radiography.* Report of AAPM Task Group #116.

American College of Radiology, *ACR-AAPM-SIIM Practice Guideline for Digital Radiography.* 2012.

Barski, Lori, CareStream Health, *personal communications.*

Bowman, Dennis, *Universal Technique Charts for DR and CR Units.* Digital Radiography Solutions. www.digitalradiographysolutions.com

Bushberg, J. T., et al., *The Essential Physics of Medical Imaging.* Philadelphia: Lippincott, Williams & Wilkins, 2011.

Bushong, Stewart. *Radiologic Science for Technologists,* 11 Ed. Maryland Heights. MO: Elsevier/Mosby, 2017.

Carlton, R., and Adler, A. *Principles of Radiographic Imaging,* 5 Ed. Cengage, 2013.

Carroll, Q. B., *Radiography in the Digital Age,* 3 Ed. Springfield, IL: Charles C Thomas, Publisher, Ltd., 2018.

Cretella, Gregg, FujiMed, *personal communications.*

Cummings, Gerald, *Producing Quality Digital Images.* Corectec@aol.com

Davidson, R. A., *Radiographic Contrast-Enhancement Masks in Digital Radiography* (Chapter 6: Current Post-Processing Methods in Digital Radiography). University of Sydney, AU, 2007. ses.library.usyd.edu.au/2123/1932/6/06Chapter5a.pdf

Fauber, T., *Radiographic Imaging and Exposure,* 5 Ed. Maryland Heights, MO: Elsevier/Mosby, 2017.

Fuji Photo Film Co., *NDT Systems.* www.fujindt.comcr_process3.html

Fuji Photo Film Co., *CR Console Operation Manual.* Tokyo, 2004.

Hearn, D., and Baker, M. P., *Computer Graphics with OpenGL.* Upper Saddle River, NJ: Pearson Prentice Hall, 2004.

Hoppner, S., et al., *Equalized Contrast Display Processing for Digital Radiography.* SPIE Medical Imaging 2002 Conference.

Jacobs, J, Pauwels, H., and Bosmans, H., *Post Processing and Display of Digital Images.* University Hospital, Leuven. www.uzleuven.be/lucmfr

Koenker, Ralph, Philips Healthcare, *personal communications.*

Kornweibel, Georg, Philips Healthcare, *personal communiczations.*

Philips Medical Systems, *Going Digital in Radiology—What You Need to Know.* Conference, Oct. 10, 2005.

Prokop, M., Neitzel, U., and Schaefer-Prokop, C. *Principles of Image Processing in Digital Chest Radiography.* Philadelphia: Lippincott Williams and Wilkins, Inc. *Journal of Thoracic Imaging, 18:*148–164, 2003.

Schaetzing, Ralph, *Agfa's MUSICA: Taking Image Processing to the Next Level.* Agfa Healthcare white paper, 2007.

Seeram, E. *Digital Radiography: An Introduction.* Clifton Park, NY: Delmar/Cengage, 2011.

Seibert, J. A., and Morin, R. L., *The Standardized Exposure Index for Digital Radiography: An Opportunity for Optimization of Radiation Dose to the Pediatric Population.* Digital Radiography Summit, St. Louis, MO, 2/4/2010.

Sprawls, Perry, *Physical Principles of Medical Imaging Online,* Resources for Learning and Teaching. www.sprawls.org/resources

SUNY Upstate Medical University: *Imaging Issues: Radiology Teaching Files.*

University of Winnipeg, GACS-7205-001 *Digital Image Processing Course:* Chapter 3 Intensity Transformations and Spatial Filtering. ion.uwinnipeg.ca/~sliao /Courses/7205/Chapter%2003.pdf

Wikibooks, *Basic Physics of Digital Radiography/The Computer.* En.wikibooks.org /wiki/Basic_Physic_of_Digital_Radiography/The_Computer.

Wolbarst, A.B., *Physics of Radiology,* E. Norwalk, CT: Appleton & Lange, 1993.

INDEX

A